COLOUR TELEVISION
SERVICING

COLOUR TELEVISION SERVICING

GORDON J. KING

T.Eng.(C.E.I.), R.Tech.Eng., F.S.C.T.E., F.I.S.T.C., F.I.P.R.E., M.A.E.S., F.S.E.R.T., M.R.T.S.

Newnes Technical Books

Newnes Technical Books
is an imprint of the Butterworth Group
which has principal offices in
London, Boston, Durban, Singapore, Sydney, Toronto, Wellington

First published 1971
Second edition 1975
  Reprinted 1976, 1979, 1983

ISBN 0 408 00464 9

Printed in Great Britain at the University Press, Cambridge

Preface

IT HAS BEEN my aim to write a comprehensive book on colour television and on the servicing of PAL receivers in the most down-to-earth manner possible and with the least amount of mathematics. The book has been written to have the maximum appeal to the technician of average qualifications changing from the servicing of monochrome sets to colour, to the student of the NTSC and PAL colour systems (bearing in mind that the latter is essentially a development of the former), to the indentured apprentice technician just commencing a career in television servicing, and to the enthusiastic amateur interested in a general way and desirous of getting the best results from his own set and those of his friends.

Colour television is an involved subject, and to be successful in the servicing of receivers one must first be conversant with the theory, servicing problems and procedures associated with monochrome models. This is because contemporary colour television is a development of monochrome television—that is, the monochrome system with additional features endowing it with colour. Service technicians who are at present familiar with monochrome servicing only will thus have new things to learn and new problems to sort out, but these will be found to dovetail into their existing knowledge, acquired during the years of servicing monochrome sets.

It would have been possible, of course, to have written this book on the basis that monochrome receivers had never existed, but then about half of it would have been concerned with colourless transmission and reception. In other words, a book of that kind would have needed to have been in two parts—monochrome and colour—and since there are many books adequately dealing with the first part, this would have meant repeating much of what has already been written on the subject.

I have concentrated, therefore, on the aspects of colour, bringing in references to the monochrome system only as and when necessary to complete the picture. This also applies to the sections dealing with servicing.

Apprentices and other readers wishing to pursue a study of television receiver servicing from first principles might find my complementary *Television Servicing Handbook* useful. This in its third edition deals with dual-standard sets, transistorised circuits and most of the recent developments in monochrome receivers. It concludes almost exactly where this book starts.

I considered it desirable to arrange this colour television book in three parts: the first describing how the PAL and the NTSC systems work, in conjunction with the science of colours; the second detailing the circuits employed in the latest hybrid and all-transistor sets; and the third concentrating on fault symptoms and servicing procedures.

PREFACE

Numerous sections are devoted to setting-up adjustments—such as static and dynamic convergence, grey-scale tracking, degaussing and purity—with manufacturers' examples, and there are chapters dealing with colour test intruments, alignment, encoding and decoding, servicing in the field and so forth, including a special chapter giving a kind of bird's eye view of the colour system as a whole, which acts as a guide to subsequent chapters.

Both dual-standard and the more recent 625-line single-standard models are examined, and a specially designed block diagram is used to show the PAL colour signal waveforms at all stages in the system, from the colour camera output to the shadowmask picture tube input. I have also prepared a block diagram to relate fault symptoms to specific areas in the set, as a simple aid to speedy fault diagnosis.

The colour signal oscillograms are real c.r.t. displays photographed in my laboratory, using standard receivers and test equipment, and some of the equipment and off-screen photographs are also of my origin; others were kindly supplied by receiver and equipment manufacturers.

Many of the illustrations and some of the early chapters describing the colour system have been restyled from my series of articles entitled 'ERT Telecolour Course', originally published by *Electrical and Radio Trading*, to which journal I wish to acknowledge my indebtedness; also to Roy Norris, its well known former Technical Editor for his great encouragement and help while writing the series and to Colin Sproxton its subsequent Technical Editor.

I should also like to thank sincerely the manufacturers and distributors of colour receivers and test equipment for answering my letters in great detail and for readily making available to me circuit diagrams and excellent photographs of their equipment, including such firms as the British Radio Corporation Ltd., Bang and Olufsen UK Ltd., Brown Brothers Ltd., Decca Radio and Television Ltd., Mullard Ltd., Philips Electrical Ltd., Pye Ltd., Rank Bush Murphy Ltd., Sony Corporation, Telefunken Ltd., Thorn-A.E.I. Applications Laboratories and all the many other firms without whose assistance this book could never have been written.

My thanks also to Marjorie Campbell for the typescript and Neil Hyslop for the excellent diagrams evolved from my rough sketches.

While the bulk of the information in this second edition remains unchanged, some updating in alignment with current practice has been considered desirable. An extra chapter has also been included to summarise some of the more recent developments in receivers.

A Fault Finding Chart based on the Procedure Charts in the *Television Servicing Handbook*, which have proved so popular, is also incorporated in this new edition.

Finally, I would like to take this opportunity to thank the thousands of readers of the first edition, and in particular those who have written to me personally through the publisher with suggestions for the second edition, some of which have been incorporated.

Brixham, Devon   Gordon J. King

Contents

CHAPTER 1

Introduction

THE COLOUR TELEVISION SYSTEM adopted by Great Britain and certain parts of Europe is called PAL, which stands for *phase alternation, line.* In the United States of America and some other countries a system known as NTSC has been in operation since 1953. This system was recommended in America by the *National Television Systems Committee*, hence the term NTSC. The PAL system is a further development of the NTSC system by the West German Telefunken Company, under the leadership of a Dr. Walter Bruch. There are other systems, but the NTSC and the PAL development of it are the main ones considered in this book.

The PAL development was considered desirable because, owing to the way in which the colour information is modulated, it is not possible to hold complete control over the colour information in the NTSC system once it leaves the cameras. As a consequence, phase changes of the colour signal can cause colour (hue) changes on the received pictures.

The PAL system effectively neutralises this at the expense of a little extra complexity at the transmitter and receiver, with the result that phase changes on alternate lines of the picture are combined in a manner that cancels out the colour error. The PAL system was chosen after being thoroughly field tested over a period of about three years in the majority of West European countries.

Colour systems can adopt time-sequential or simultaneous methods of colour processing. The sequential method means that the colour information is presented to the picture tube as little 'bits' during specific periods of time which are geared to the field, line or subcarrier frequency. These are called respectively field-sequential, line-sequential and dot-sequential. Colour synchronising is necessary here to ensure that the reproduced 'bits' of colour information occur at the receiver at exactly the same instant as they are produced at the transmitter, from the colour camera.

In the USA a field-sequential system was commissioned by the FCC (Federal Communications Commission) in 1950, but was revoked in 1953. It was semi-mechanical, using three colour filters rotating at field frequency in front of the monochrome picture tube and was unsuccessful commercially, if not technically. Colour systems were also tried out using in the UK line-rate and switching periods which were small compared with picture elements (i.e. dot-sequential).

The NTSC system and its PAL development, however, use the simultaneous method, where monochrome and colour picture element information are presented simultaneously to the display tube.

All colour systems must be *compatible*. This means that the transmitted colour-encoded signals must produce monochrome pictures of good quality on ordinary black-and-white sets, adjusted to suit the line standard and programme channel. They must also have *reverse compatibility* to allow colour sets to reproduce in monochrome from a monochrome signal. Both NTSC and PAL have these attributes.

The difference between PAL and NTSC is worth a basic examination. The chief difference is in signal handling at the receiver and transmitter to combat phase errors and hence incorrect colour presentation.

Colour can be fully described by three parameters—hue (i.e. the actual colour), saturation and luminance. The luminance in colour systems is handled in the same way as the monochrome information in black-and-white sets. Hue and saturation are handled by separate colour signals in the system. It is shown in later chapters that the information on hue appears as a phase-modulated component, while that on saturation appears as an amplitude-modulated component. This means that *amplitude* changes of the signals simply affect the saturation of the colour, for instance, causing a red rose to appear either a deeper red (increasing amplitude) or less red, towards pink (decreasing amplitude). Phase change, however, is more of a problem. Much more, in fact, since it can change green grass into red grass!

These components of the colour signal can change in both amplitude and phase anywhere between the television camera and the picture tube due to a diversity of factors at the studio, the transmitter, the receiver and even in space. Amplitude changes can be tolerated, but even the smallest phase change is immediately discerned as a distressing change in colour.

NTSC sets feature a *hue control*, which allows adjustment of the colour phase at the set by the viewer to correct for system phase error. It is the viewer's job, therefore, to adjust hue on a colour test card or for the most natural reproduction of flesh tones—making his own judgement as to what he thinks is the correct reproduction. This can cause problems; but with PAL the colours are always presented correctly once the receiver circuits have been properly adjusted. The only colour control that the PAL user has to bother with is that for setting the saturation. Indeed, with this control turned right down the picture is reproduced in black-and-white, and colour intensity is progressively increased as the control is turned up. This is the *saturation control*.

Some PAL sets do have a further colour control, labelled *tint control*. This can be adjusted on a black-and-white picture to secure the most desirable monochrome colour temperature. We shall see later that white on a colour tube is given by red, green and blue lights superimposed, and the nature of the white is governed by the balance of these three coloured lights. The tint control, therefore, simply allows the red, green and blue lights to be adjusted differentially to give the white most preferred by the viewer. The control can also be used to compensate for unbalance in the tube and video supply circuits.

PAL has developed over several stages. There is, in fact, a simple PAL system (PAL-S). and a de luxe version (PAL-D). These titles refer to the receivers, not to the transmission. That is, PAL-S and PAL-D receivers will work from the common PAL transmission. The PAL transmitter switches the phase of one colour signal over 180 degrees line by line. This will be difficult to appreciate now, but the technique will become apparent.

If there is a wrong hue caused by system phase distortion, the line-to-line phase switching will present the incorrect hue on one line and the complementary hue to this incorrect one on the next line of the particular field. The subjective effect of this is that the viewer sees the error and its complement together, and the impression is then given of the true colour. This is on PAL-S sets.

PAL-D sets go a stage further than this, for the line-by-line opposing errors are fed into an electronic circuit, averaged, and the electrical output presented to the tube in corresponding true colour. Here the eyes are not called upon to assist with the error averaging—it is done electronically. PAL-D sets can cope with errors up to about 70 degrees, while PAL-S sets can cater only for smaller errors.

PAL-S sets have a shortcoming, however, for when the error exceeds a few degrees the eyes see the two lines of opposing colours separately. This is aggravated by the interlaced scanning, since this puts together two lines in colour error and two lines in colour error correction successively over the whole picture. The viewer then becomes aware of horizontal lines or bars of colour across the picture, which in colour television are referred to as 'Hanover blinds' (see Chapter 5).

PAL-D sets use an electrical delay line which effectively stores the colour information of one line for its duration (about 64 µs) and then delivers it during the next line of signal along with the colour information of that line, but in opposite colour phase error. The error is thus cancelled out electronically, and the Hanover blind effect is avoided.

A by-product of phase error compensated by PAL-D is a change in colour signal *amplitude*, but the phase error in PAL-D sets needs to be substantial before saturation changes become apparent; but this is all that can happen, and it is far less disturbing than changes in hue.

The process of adding colour signal to the monochrome signal at the transmitter is called *encoding*, and extracting the colour signal from the monochrome signal at the receiver is called *decoding*. A receiver's decoding system must, of course, match the nature of encoding. Many encoding/decoding systems have been evolved over the years and patented, including the French SECAM, the Soviet NIIR, the German FAM and others. The components of the signals derived from the primary colours of the SECAM system are transmitted sequentially, as already explained.

PAL sets feature many of the familiar 625-line standard black-and-white circuits, including the intercarrier sound channel, the i.f. strip, u.h.f. tuner and so forth. Indeed, a colour set can be considered as a monochrome set of 625-line or dual-standard—with refinements to the tuned circuits, power supplies and timebases—plus two main colour sections. These are the electronics to process the colour information (decoder and colour channels), and the colour display (shadowmask tube) with its controlling and correcting circuits.

The monochrome parts of the set produce the *luminance* signal, which is the same as the signal delivered by black-and-white sets, while the colour electronics deliver red, green and blue primary colour signals required by the colour tube. The elementary arrangement is shown in Fig. 1.1. The composite signal carries all the information necessary for a black-and-white picture plus the colour

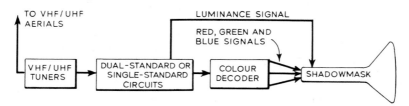

FIG. 1.1. *Elementary diagram of a colour receiver, showing the basic additions a monochrome set needs for colour.*

information which is used only by colour sets. Thus, monochrome sets working from a colour transmission by-pass the colour information.

At the transmitter, the signals from the colour camera are first processed into *colour-difference signals*, which are then specially modulated, giving chrominance signal, and sandwiched into the luminance signal. To facilitate extraction of the colour signals at the receiver, bursts of colour subcarrier are developed on the 'back porch' periods of the line sync pulses, as shown in Fig. 1.2. These are called *colour bursts*, and their purpose is to help recreate a subcarrier at the

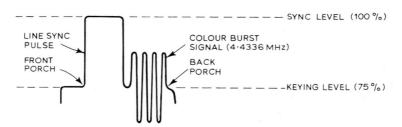

FIG. 1.2. *Colour bursts contain 10 cycles of subcarrier signal on the back porches to the line sync pulses. These signals lock the locally generated subcarrier signal at the receiver to the phase (and hence frequency) of the subcarrier upon which the chrominance information is modulated at the transmitter, but not transmitted.*

receiver which is identical to that at the transmitter on which the colour signals were originally modulated, but which was not transmitted. The bursts also identify the phase alternations of the PAL lines.

The colour-difference signals represent the difference between red primary and luminance signals and blue primary and luminance signals. The green colour-difference signal is not transmitted, but is recreated at the receiver. These techniques help towards bandwidth saving and ensures that when there is no colour in a televised scene there is no colour signal.

4

INTRODUCTION

The shadowmask colour tube is normally used with current television receivers and monitors. In this tube, for any particular spot or element of picture, there are three minute dots of screen phosphor which glow red, green and blue when bombarded by electrons. There are three electron beams, and each one is focused upon its own corresponding colour dot. The phosphor dots are not coloured, just the light they emit when bombarded. At viewing distances the eye is aware of only a single spot consisting of a mix of the three colours.

The mix is changed by *relative* changes in the strength of the three beams, and for white light the light outputs of the three colours have to be balanced. The intensity of white light is thus varied by the three beams changing strength equally and in the proportions required for white light. This gives all the shades of grey, from peak white down to black (beam cut-off).

Although the shadowmask tube looks almost the same as its monochrome counterpart from the outside, the inside is much more complicated, as shown in Fig. 1.3. Figure 1.4 shows the outside appearance. Colour tubes are much

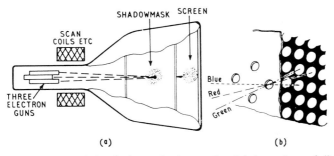

FIG. 1.3. *The basic features of the shadowmask picture tube, (a) in section and (b) part of the shadowmask.*

more expensive than similar size monochrome tubes. Fortunately colour tubes appear to be longer-lasting than monochrome ones, and in the USA some tubes are still giving good pictures after 13 years' use.

Colour tubes need a greater e.h.t. potential than monochrome ones, with a final anode voltage often of the order of 25 kV. This means that soft x-rays are produced by the resulting high velocity electrons, and precautions are taken in tube manufacture to prevent them from harming the viewer. X-rays can also occur inside the set from the e.h.t. rectifier and regulator valves, but internal screening and the use of semiconductor devices reduce troubles of this kind. It is most important, therefore, that all screens and shields are correctly replaced after servicing operations. Sets using semiconductors throughout are less prone to internal x-rays, because these devices do not generate them in the same way as valves.

The three beams of the shadowmask tube have to register accurately with their colour dots to ensure the right colour mix. This is also necessary to obtain good-black-and-white reproduction on a colour tube. Colour fringing effects are soon apparent if the registration is wrong. Registration is achieved by beam convergence correction of two kinds—*static convergence* and *dynamic convergence*. The first uses fixed magnets on the tube neck which line up the three beams in the middle of the screen. Dynamic correction is achieved by passing corrective currents through windings on the neck-mounted convergence unit. These currents are obtained from the line and field timebases, and they preserve the registration when the three beams are deflected away from the centre of the screen.

It will be appreciated that the colour tube is sensitive to magnetic fields, and stray fields near the neck or flare will impair the beam convergence. Stray fields

FIG. 1.4. *External appearance of shadow-mask picture tube (courtesy, Mullard Ltd.).*

can result from the magnetisation of steel parts within the set and shadowmask tube, and it is essential that all such fields are neutralised before the tube is adjusted for *purity* (see below).

This is achieved by a process of degaussing (meaning demagnetising) based on 'influencing' the set, tube and metal parts with the field from a large coil of wire connected across the a.c. mains supply. All sets now have a degaussing coil built into them, which comes into operation each time the set is switched on. This is called *auto-degaussing*.

While convergence control affects the three beams separately, another neck-mounted magnet, called the *purity magnet*, affects the three beams together. This ensures that the three beams approach the shadowmask at the correct angles so that they pass through the right holes to strike their appropriate colour phosphors in the centre. If this does not happen (tube not pure), the beams may strike the edges of the phosphors and cause wrong colour dots nearby also to emit light. This is corrected by the purity magnet.

In Great Britain colour sets are made both as single 625-line standard and dual-standard models. Only the u.h.f. channels, however, carry the colour signals of the BBC and ITV. The plan is for all 405-line standard programmes to be duplicated on the u.h.f. channels and then at some time in the future for the main v.h.f. transmissions on 405 lines to cease. By then the majority of viewers will be in range of a u.h.f. regional group (carrying four u.h.f. channels), and the old v.h.f. channels may then be re-engineered to take 625-line standard programmes, on monochrome and colour, to cater for those areas which are badly shielded from the u.h.f. signals. It will take at least a decade to put all this fully into operation.

In the meantime, dual-standard colour (and monochrome) sets will be continued, but with more sets being made for the 625-line standard only, for those areas in which all programmes can be received on the u.h.f. channels.

It is important that colour sets are fed with an adequate aerial signal, for a weak signal can produce an abundance of coloured grain (noise). U.H.F. signals are more difficult to receive adequately than v.h.f. and, since colour is in the u.h.f. channels (to start with, anyway), this is another good reason why special care must be taken over the aerial installation feeding a colour set.

Colour sets will work from communal aerials and relay systems, but poor results can be expected on those systems in which phase and amplitude distortion (and non-linearity) are present. These systems work by distributing over the cable u.h.f. signals at natural frequencies or at v.h.f. (or lower frequencies)— the latter using a u.h.f.–v.h.f. converter at the aerial end. These, however, do call for alterations in the colour set so that it will work on the 625-line standard when the channel selector is adjusted to a v.h.f. channel.

CHAPTER 2

The Science of Colours

THE SOURCE OF the Earth's light and heat is the sun, which emits a wide range of radiation, from ultra-violet through the visible light spectrum into the infra-red and heat radiation wavelengths. This radiation is in the form of electro-magnetic waves and comprises a very small part of the overall electromagnetic waveband, which includes the waves we use for radio and television trans-missions and other more dangerous waves such as x-rays, and gamma-rays which are caused by the release of nuclear energy. The only difference between radio and light waves, and heat and other waves from the Sun, is that of wavelength.

LIGHT WAVES

Light and heat waves are close together in the scale of radiations but substan-tially removed from the wavelength of radio waves. Radio waves usually have their wavelength measured in metres (m), while light and heat waves are so short that they are best measured in micrometres (μm), millimicrometres (mμm)* or Ångström units (Å). A micrometre is a millionth of a metre (10^{-6} m) or a thousandth of a millimetre (10^{-3} mm). A millimicrometre is thus a thousandth of a micrometre (10^{-9} m or 10^{-6} mm). Colour television sometimes makes reference to the Ångström unit which is 10^{-10} m or a tenth of a mμm. The reason for using these units is that they enable light wavelengths to be considered in whole numbers. Each wavelength has a corresponding frequency, given by dividing its value in metres into the velocity of light, which is 300×10^6 m/s. A wave 1 mμm in length, for instance, has a frequency of 10^{16} Hz. Since light wavelengths are in hundreds of millimicrometres, however, the frequencies fall in the 10^{14} Hz range.

Figure 2.1 shows the overall electromagnetic wave spectrum with the visible sunlight band ranging from red to violet in hue and from 700 to 400 mμm in wavelength. The frequency range, which is not often used to describe colours, ranges from about 3.8×10^{14} to 7.9×10^{14} Hz, the violet end having the highest frequency and the shortest wavelength.

Although sunlight consists of this spectrum of colours or hues, the eye registers them collectively as white light. It is possible, however, to 'tune' the eye, so to speak, through the light spectrum by using a special optical instrument

---

* Often expressed mμ. The nanometre (10^{-9} m) is also gaining popularity.

or, more easily, to select certain wavelengths passing to the eye by attenuating the unwanted ones. A piece of red glass plate, for instance, is a red-pass filter when placed in front of the eye, since it lets through only waves of the red wavelength from the sun or other light source while greatly attenuating the remaining waves.

White light is usually defined as that given by the noon sun on a clear day. This light, as we have seen, is a band of radiations and is called pure. It ceases

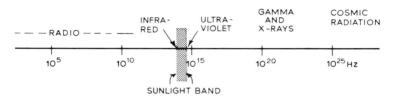

FIG. 2.1. *The electromagnetic wave spectrum.*

to be pure when some component radiations are missing or attenuated. Morning and late afternoon sunlight is not pure in this context—nor is the light on a dull day. This is why colour photographs fail to reproduce colours correctly when taken during these times without a special lens filter.

Artificial light is not pure, but approaches purity when it is emitted from arc sources. A great deal of work has been undertaken by lighting engineers and others in this connection, but pure light is not necessarily what the public wants from lamps or from television screens for that matter.

COLOUR SPECTRUM

One way of analysing light is to pass it through a prism or train of prisms. The emerging light is then dispersed in colour bands to form a spectrum display. This is caused by a process known as refraction. As light passes from air to glass, and vice versa, its direction of travel is bent, due to the different refractive indices of air and glass. The higher the frequency, the greater the degree of bending. Figure. 2.2 (left-hand side) shows how white light passed through a prism is dispersed into spectrum hues. By passing the spectrum colours through a second prism, or through a lens, reciprocal refraction occurs and the original white light re-appears as shown by the right-hand side of Fig. 2.2. Naturally, if one or two of the colours are removed or altered in energy before recombination, the original nature of white light will not be obtained.

However, it is this phenomenon that makes colour television possible, and if just three colours—or even two in some cases—remain the eye is deceived into seeing white. Indeed, we can get not only white but a great range of hues by varying the proportions of the mix of lights. With two colours only, the range of colours obtainable is limited.

Let us suppose that three projectors are set up in such a way that three overlapping pools of light are thrown upon a white screen, and that each projector has a colour filter placed in front of it. Given a wide range of filters of various colours and plenty of time to experiment, it would eventually be discovered that

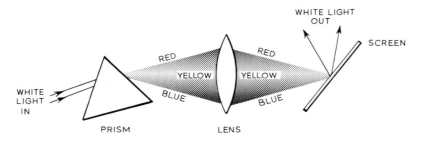

FIG. 2.2. *The colour spectrum obtained by passing white light through a prism can be recombined by reciprocal refraction to give the original white light, as shown here.*

white light would appear when one projector has a red, the second a green and the third a blue filter placed in front of its lens, as shown in Fig. 2.3.

It would be necessary to experiment with the energy of light from each projector. White would appear only when the energy of the projected light from each projector was of a 'standard' value. This value would occur for a particular kind of white characterised by the filters used.

ADDITIVE PRIMARIES

Red (R), green (G) and blue (B) are called additive primary colours, and the standard value of each to give white light when the coloured lights are added

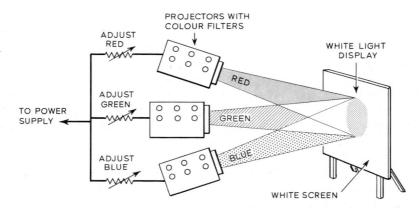

FIG. 2.3. *A three-colour projector can produce white, greys and colour displays (see text).*

together is called a 'primary unit' (pu). Thus, 1 Rpu + 1 Gpu + 1 Bpu = white. If the light output from each projector is increased equally, the colour additions will give a brighter white, while a duller white will occur if decreased equally. In other words, it is possible to swing from black (projectors turned off) up to maximum white (all projectors running at maximum). Between black and the maximum possible white energy that the system can deliver is the grey range. The degree of grey, which can only be a less-white white, is relative to the maximum white and to black. When a white shirt, for instance, is viewed on a monochrome television set with a worn-out picture tube, it still seems white to us because this particular white is the maximum white that the exhausted tube can provide. White is thus relative and white and greys are colourless or achromatic.

In addition to providing white and a whole range of less-whites (greys), the three-projector system will give *red* (blue and green projectors out), *green* (red and blue projectors out) and *blue* (red and green projectors out). These are displays of the basic additive primaries with no colour mix. If two projectors are running we get another three colours. Red and blue produce a purplish hue called *magenta*. Red and green give a bright *yellow*. And blue and green produce a bluish-green called *cyan*.

So far, then, it has been shown how all the greys plus six different colours can be obtained. But this is only a start, as almost all the spectrum colours can be obtained by light mixing, as we shall see.

There is an important point about the mixing so far described, for artists know that when they mix blue and yellow pigments they get green—certainly not yellow by mixing red and green! The reason for this is not difficult to understand. It has to do with the difference between mixing *lights* and mixing *pigments*, paints, crayons etc. When lights are mixed, the eye *adds* the colours, but when pigments are mixed, the eye discerns them subtractively.

SUBTRACTIVE PRIMARIES

Artists' primary colours are not really red and blue, although they are often called these colours. In fact they are magenta and cyan, which together make yellow. In other words, the colours used by artists are those obtained with two projectors running. These are called *subtractive primaries*.

Consider magenta again: this is obtained by running the red and blue projectors with the green one out. It is the same as having white and subtracting green. Similarly, white minus red gives cyan and white minus blue gives yellow. Let us suppose now that we mix a pigment of cyan and yellow and that this mix is viewed in white light. The cyan (white minus red, remember) absorbs red from the white light and leaves blue and green. However, the yellow of the mix (white minus blue) absorbs the blue, leaving the reflected green seen by the eye.

Four other examples of subtractive mixing are given in Fig. 2.4. At (*a*) the yellow pigment absorbs blue and reflects red and green together; adding at the eye gives the sensation of yellow. The magenta pigment at (*b*) absorbs green and reflects red and blue; adding at the eye gives magenta. At (*c*) the cyan pigment absorbs red and reflects green and blue, giving the sensation of cyan. Finally, at (*d*) the mix is yellow, magenta and cyan which, all being subtractive primaries, absorb all the additive primaries from the sun's light, leaving zero reflection or black.

To sum up, a black object reflects no light to the eyes and remains black; a white object reflects all the colours and thus assumes the colour of the light falling upon it; while a coloured object reflects only the waves of the light corresponding to its particular colour. It is important to note that colour is subjective and that the addition of *light* takes place in the eye itself. It is not intended to delve deeply into the manner in which the human eye works, but it is worth noting that the lens of the eye focuses an image of the scene on the retina, and that the

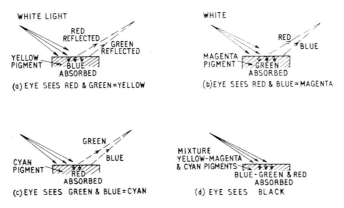

FIG. 2.4. *Examples of subtractive mixing.*

retina is composed of light-sensitive organs which generate impulses corresponding to the information of the image. These impulses are fed to the brain and inform it of the scene.

There are two kinds of light-sensitive organ in the eye, called respectively *rods*, which provide the high-definition monochrome (detail) vision, and *cones*, which provide mostly colour vision. There are about ten times as many rods as there are cones and this is one reason why we can define an object minutely in white light but not in coloured light.

Indeed, it is a good thing for colour television that the eye is so accommodating, for it is this characteristic that enables colour television to work on the basis of high definition monochrome (giving the detail) with relatively low definition colour added.

COMPLEMENTARY HUES

Subtractive primaries (yellow, cyan and magenta) are called *complementary colours*, and each one is the complement of the additive primary that was subtracted from white light to produce it. The respective complementaries of red, green and blue are thus cyan, magenta and yellow. By adding the additive primary to its complementary colour, white is produced. For example, yellow added to blue gives white. This is because yellow was obtained by subtracting blue from white—the blue is simply put back again. Similarly, cyan added to red gives white, as does magenta added to green.

Almost any colour can be produced by adding different amounts of red, green and blue light. It is difficult to visualise the wide range of hues that can be obtained from these three primaries by changing their relative energies (intensities), but the chromaticity diagram helps in this respect. This is based on an internationally agreed way of presenting colour and it has been in use now for over 36 years, so it is by no means a colour television invention!

The chromaticity diagram is shown in Plate 1. The horseshoe shape contains all the colours in nature, with the highly saturated colours occurring away from the centre-point marked C. The locus of the horseshoe shape is calibrated in millimicrometres (wavelength). Towards the centre the saturation reduces up to the achromatic area, where there is no colour.

Complementary colours lie where a line, passing through C, cuts the locus at two points. The main ones, as we have already seen, are yellow, cyan and magenta. Others can easily be found on the diagram. While the diagram is based on spectral hues, dot phosphors of colour picture tubes cannot correspond exactly to these, and this restricts the range of colours available to a small degree. It is possible to see how the colours are limited relative to the full spectrum (given only by the sun) by plotting at the appropriate points the light wavelengths radiated by colour phosphors used on colour tube screens, and then joining the points by lines to form a triangle within the colour horseshoe.

TELEVISION COLOURS

The colours produced on a TV screen using phosphors of the colour characteristics plotted are confined to those colours within the triangle. This is of small consequence because of the accommodating nature of the eye. Only the highly saturated primaries are affected, as can be seen. The wavelengths of the phosphors in colour TV tubes are about 610 mμm red, 535 mμm green and 470 mμm blue. These points correspond to the corners of the triangle within the colour locus.

Light can be analysed into three components—hue (wavelength), saturation (the degree by which the colour has been diluted by white light) and luminance or brightness. The distinction between saturation and brightness should be appreciated. In the colour projection system, for instance, the saturation of any one projected colour remains constant at all levels of projected intensity. The colour thus changes in *brightness*, but not in saturation.

Saturation is changed, however, by the addition of white light to the colour. That is, by the addition of lights corresponding to the three primary colours. For example, a deep red changes to pale red, through pink and eventually towards white as its saturation is decreased by the addition of white.

We have seen that the three groups of cones of the eye respond to the three primary colours. However, each group is not sharply 'tuned' but responds to a band of frequencies, and the three bands overlap to give the staggered response effect of some turned r.f. circuits. This is shown in Fig. 2.5.

When the eye responds to light—other than one primary colour, which is rarely—all three types of cones send impulses to the brain, thereby producing a subjective awareness of the actual colour of the light. This happens continuously for all the many thousands of 'elements' composing a scene in colour.

The three curves in Fig. 2.5 indicate the 'sensitivity' of the eye to all colours, with peaks at red, green and blue. The diagram is not meant to show the relative subjective sensitivities to red, green and blue. This is impossible because all human eyes are not 'matched'. A colour-blind person, for instance, may have a low red or blue response; and it is possible for a person to be totally colour blind to one of the primary colours, and less frequently to two of them.

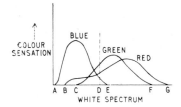

FIG. 2.5. *Colour response of the human eye.*

The horizontal axis of Fig. 2.5 corresponds to the overall spectrum of white light. The blue curve extends from *A* to *E*, the green from *C* to *F* and the red from *B* to *G*. The vertical axis refers to the 'sensation' of colours. From *A* to *B* the sensation is of blue (going towards violet), from *F* to *G* pure red; but from *B* to *F* the three colour sensations are mixed. This means that *pure* green fails to exist by itself in the eye response. Green is thus seen at point *D* where the eye gives a mix of red, green and blue sensations.

The colour cones are stimulated together to give the sensation of colour, as it seems to us individually. A colour-blind person can usually see colours, but differently from those without the affliction. A red-blind person, for example, would fail to have full use of the red curve, so that his colour sensations would result mainly from mixes of blue and green, and it is also possible that the responses to these colours would be different—larger or smaller, from those of a person with normal colour vision.

Measures of the response of the eyes to colours demand the use of special dot-pattern screens, pictures and more scientific devices, outside the scope of this book. However, it is of academic interest to know that almost 10 per cent of the male population is colour-blind. Such persons recognise colours, and can name them, but apparently cannot see colours as brightly as persons with normal vision. This percentage of colour-blind service technicians has nothing at all to worry about. There will be no undue difficulty in setting up PAL colour sets and enjoying the pictures. However, there is a small percentage of people who are distinctly blind to one or more (sometimes all) primary colours. In bad cases these people cannot see colours at all, so it is obvious that people with this affliction would be unable to practise as colour television service technicians.

Measurement of the response of the eye is based on a subjective unit called the *lumen*. This is not the same as the energy of light radiation or primary units which were considered earlier. An understanding of this is possible by considering two lights of different colours. If these are registered by the eye as equally bright, then they would have equal lumen value. However, because of the eye's

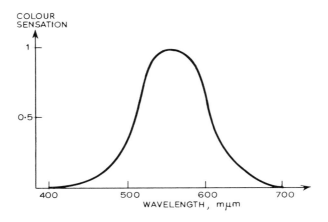

FIG. 2.6. *The luminosity curve, obtained for an 'average eye' by the integration of the curves in Fig. 2.5. Note maximum colour sensation occurring between green and yellow.*

dissimilar responses to colours, *energy* radiated at each colour would also need to be dissimilar to give equal brightness impression.

Integration of the curves in Fig. 2.5 gives an overall response curve, called the *luminosity curve*, shown in Fig. 2.6. One might query how this can be drawn in so definite a manner when the responses of eyes differ from person to person. The answer is that the curve is based on a theoretical concept of a so-called 'standard eye', the parameters of which were devised many years ago by the C.I.E. (International Commission on Illumination). The curve rises to a maximum in the yellow-green part of the spectrum at about 550 mμm, corresponding to the maximum energy band of sunlight.

EQUAL-ENERGY WHITE LIGHT

Equal-energy white light is produced when the energy of radiation is the same for each colour. 'Standard white', called *illuminant C*, is (also *see illuminant D*, below) described as the average colour of the north sky—and corresponding to point C in Plate 1—is not truly equal-energy white light, for its response rises a little towards the blue part of the spectrum. This light, and the light of the Sun about mid-day are compared with equal-energy white light in Fig. 2.7. Noon sunlight tends to *fall* towards the blue end of the spectrum, while illuminant C rises here. This is why a 'white' raster produced by a colour set tends to appear a little on the blue side, which can also be said of some monochrome television pictures.

Some colour sets feature a control labelled 'personal tint' (or a similar name) which allows the raster of a monochrome picture to be adjusted from bluish-white, through illuminant C (and illuminant D) to sepia white, as the viewer

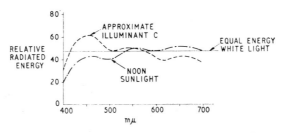

FIG. 2.7. *The 'white' light from a television screen and noon sunlight are here compared with equal energy white light.*

may prefer. We shall see later that this control can also serve to adjust-out tracking errors between the three electron guns of the shadowmask tube which may develop after a period of use.

It is noteworthy that since writing the main text, the BBC has adopted a slightly different, 'warmer' white reference called *illuminant D*. A colour is sometimes referred to as a 'colour temperature', and in this connection illuminant D is the white light produced when a solid is heated to a temperature of about 6,500°C.

The luminosity curve suggests that the green-yellow part of the spectrum is most important for conveying brightness sensation. This is, in fact, true of colour television as the colour signal for green is the largest and that for blue is the smallest, as we shall see.

The Colour Camera, Signals and Displays

ONE KIND OF colour TV camera uses three camera tubes with colour filters interposed between them and the scene so that they respond separately to red, green and blue lights. An image of the scene is focused on the photoconductive surface of each tube, which is then 'scanned' by an electron beam to generate the video signal. The resulting three video signals are eventually combined so that one camera output corresponds to the monochrome composition of the scene. This is called the *luminance signal* (symbol Y) and it is this combined signal that makes black-and-white reception possible within a colour system.

The scheme clearly calls for the three camera tubes to operate with images coinciding exactly in the same field of view, focus etc.; and also with electron scans that are identical in dimensions and linearity. Any unbalance, optical or electronic, will degrade the luminance signal and hence the monochrome picture.

Partly to ease these problems, some colour cameras incorporate four camera tubes, one of which is for the luminance signal alone. For instance, because the

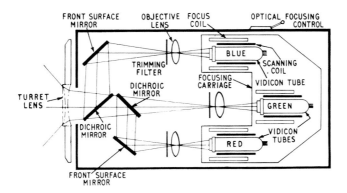

FIG. 3.1. *Basic elements of a three-tube colour television camera.*

colour signals do not require fine detail (Chapter 2) the three tubes, upon which the colour images are focused, are not so critical to set up for accurate registration.

Whether the Y signal is obtained from a colour signal mix or from a separate (fourth) camera tube, it has the same relationship to the red, green and blue outputs. The optical layout of a typical three-tube camera is shown in Fig. 3.1.

This is designed to allow the maximum amount of light to fall upon the photo-electric surfaces of the camera tubes by means of dichroic mirrors.

DICHROIC MIRROR

A dichroic mirror passes light of all colours except one selected colour which it reflects. The idea is illustrated in Fig. 3.2. Here the mirror passes light of all colours except blue, which it reflects. If a second dichroic mirror is arranged to pass only the green light of the two primary-colour lights passing through the first mirror and to reflect, say, the red light, the scene is then analysed into the three primary colours. Blue is reflected by the first mirror, red by the second, while green light passes through the system. In Fig. 3.1, light entering the camera through the lens turret is split into the three primary colours in this way. Two dichroic mirrors subtract first the blue light and then the red light from the white

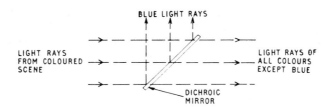

FIG. 3.2. *How a dichroic mirror works. One primary colour is reflected by the mirror and the other two pass through it.*

light input, and both pass the green light. The three lights are focused on to the photosensitive surfaces of the camera tubes by objective lenses. These are designed to have a small field of view to avoid the colour errors which would otherwise be introduced when dichroic mirrors are used at widely different angles of incidence.

Trimming filters in front of the objective lenses provide the camera tubes with wide spectral responses, like the three primary colour responses of the eye, so that most light inputs produce signals from all three tubes.

The video outputs of the three tubes are generally tailored to be of equal voltage when the camera is scanning a pure white scene. The three outputs of a three-tube camera are added in a special way to give the Y signal in addition to separate red, green and blue signals.

DERIVING THE Y SIGNAL

The requirements for a correctly balanced Y signal are 30 per cent red signal, 59 per cent green signal and 11 per cent blue signal. Whether the camera has three tubes or four, the Y signal is always equal to $0.3\,R + 0.59\,G + 0.11\,B$, where R, G and B are the primary-colour signals from the appropriately filtered tubes. This means that a *monochrome* camera, initially adjusted to provide a video output of 1 V when scanning pure white, would give 0·3 V when scanning saturated red, 0·59 V from saturated green and 0·11 V from saturated blue (Fig. 3.3). *This is because the Y signal (the monochrome signal) from a colour camera is similar in character to the signal from a monochrome camera.*

Accurate potential dividers are used to obtain the correct colour signal ratios for the Y signal, as shown in Fig. 3.3. Here it is assumed that, when scanning a white scene, each camera tube is set to deliver a 1 V video signal. Thus, the outputs are 1 V red, 1 V green and 1 V blue. The potential dividers add these three signal voltages in the correct proportions to produce the Y signal, which also has a level of 1 V. When a separate luminance tube is used in the camera, its *white* output would also be 1 V.

Let us consider the four-tube camera. Light from the scene is split into red, green and blue components by prism-type dichroic surfaces (which function in a similar way to dichroic mirrors), and the three colour components pass through colour-trimming filters (as in Fig. 3.1) and a system of lenses to the three tubes.

There is also the luminance tube and, when the camera is working mono-chrome, a fully reflecting surface is shifted into position in place of the colour

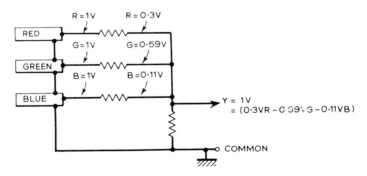

FIG. 3.3. *Each tube of a colour camera is adjusted to a reference video output level, which is 1V in this diagram. The signals are proportionally attenuated to give red 0·3V, green 0·59V and blue 0·11V, and when added together the Y signal is also at 1V level.*

light-splitting prism, thereby passing all incoming light to the luminance tube alone. Three- and four-tube colour cameras now employ vidicon photo-conductive tubes instead of the image orthicon tubes, which were previously adopted in some three-tube cameras.

It will be observed that the four-tube camera seems to be a closer approach to the human eye than its three-tube counterpart for it will be recalled (Chapter 2) that the eye also has *four* groups of organs, three for the primary colours (cones) and one group mainly for the brightness (or luminance) called rods.

Diagrams (*a*) to (*g*) in Fig. 3.4 show a supposed arrangement of colour cameras scanning screens of pure white and of the primary and complementary colours. The seven screens, therefore, are white, red, green and blue (primary colours), and yellow, cyan and magenta (complementary colours). It will be recalled from Chapter 2 that white is produced when a complementary colour is added to its primary.

Diagram (*a*) shows that when a colour camera is set up, as in Fig. 3.4(a), to scan a white screen, the Y output is 1 V. The Y output is changed from unity when the whiteness of the scene or its illumination changes. The red, green and blue outputs also change, of course, but the *proportions* hold steady. It is with this unity Y signal setting of the camera that the other screens—of full saturation (meaning no white mix)—are scanned in turn.

With the red screen (*b*), only the red tube delivers signal. This is equal to 0·3 V at the Y output. With the yellow screen (*c*), there is zero blue output, but outputs from the red and green tubes. These are full outputs because red and green in equal primary units produce yellow. Thus, the red tube contributes 0·3 V and the green one 0·59 V to the Y signal, giving 0·89 V. With the green screen (*d*), the green tube only contributes to the Y signal, giving 0·59 V. With the cyan screen (*e*), the blue and green tubes contribute signals of 0·11 V and 0·59 V respectively, adding to give the Y signal of 0·7 V. This is because blue and green in equal primary units produce cyan. With the blue screen (*f*), the Y signal output comes from the blue tube alone, because blue is a primary colour, giving 0·11 V. With the magenta screen (*g*), red and blue outputs of 0·3 V and 0·11 V respectively contribute to the Y signal, giving it a value of 0·41 V.

The camera thus analyses all colours in this manner, and the Y signal is always the sum of the *proportioned outputs* from the three tubes. Figure 3.5 shows that yellow and cyan produce the brightest sensations, with green not far short. Colours are thus produced as shades of grey on monochrome sets; relative to white, yellow is very light grey, cyan is light grey, green is mid-grey, magenta is mid-to-dark grey, red is dark grey and blue is very dark grey.

Having now seen how the colour television camera splits the light reflected from the scene into the three primary colours and produces four signals—one for each primary colour and one for luminance (Y)—let us see how these signals can be translated into monochrome and colour displays.

Picture displays on colour sets are given on the shadowmask tube—the tube almost exclusively adopted at the time of writing throughout the world for domestic colour television. This tube is fully dealt with in Chapter 5, but an intermediate step towards its understanding, taking in the historical evolvement of colour television, is provided by a system using three separate display tubes. Indeed, the idea was active in 'prototype' domestic sets years ago and, in spite of bulk and other drawbacks, it has been (and still is) employed in certain closed-circuit colour systems.

Let us first recapitulate on the basic aspects of the colour signals and the requirements of compatibility. When a colour camera televises a scene devoid

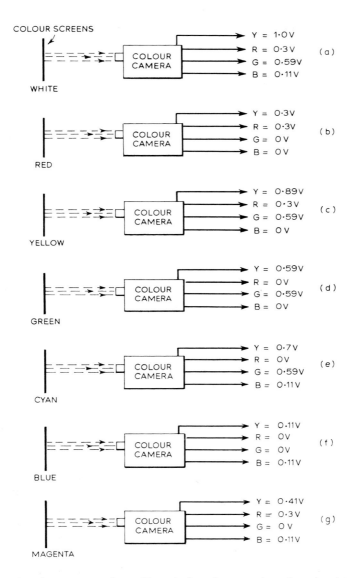

FIG. 3.4. *How the colour camera analyses white and coloured scenes and translates them into luminance (Y) and colour signals. Notice that Y in all cases is equal to the sum of the signals from the three tubes.*

of colour, what are the signals it produces? The scene will reflect some of the white light of the studio and this, as we know, consists of all the spectral colours. Thus, the red, green and blue tubes in the camera will deliver video signals in the proportions red 30 per cent, green 59 per cent and blue 11 per cent.

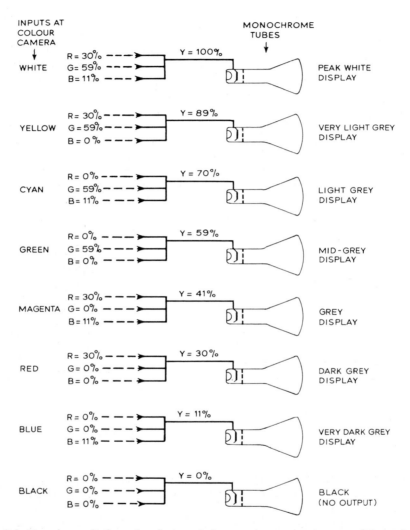

FIG. 3.5. *Monochrome displays when the input is from a colour camera in terms of Y signal. The composition of the Y signal is also shown in terms of colour signal levels based on 100 per cent colour saturation.*

Whether the camera uses only three tubes or includes a fourth for the Y signal, the output circuits will deliver a luminance (Y) signal, equal to the sum of the R, G and B in the above proportions. Any signal in these proportions can thus be described as '100 per cent signal'. Note that this does *not* refer to actual signal level.

Where the monochrome scene is most bright, the 100 per cent will coincide with the peak white signal; but at less bright parts of the scene the 100 per cent will correspond to lower-level signals. The important aspect here is that as long as the scene is devoid of colour, the R, G and B primary signals remain in the Y proportions.

The above thus considers the camera scanning a colourless scene. Now let us suppose that the camera is directed to a coloured scene. Of course, if there happens to be an object in the scene reflecting peak white light, the Y output will again be 100 per cent relative to the R, G and B primary signals, and also at peak level. But from every other part of the scene, however bright the colours, some of the colour components must be less than their 'white' values and so the sum—the Y signal value—must be less than 100 per cent.

Put in another way, all colours are represented by a whole scale of Y signal values. Because the colour signals comprising the Y signal are geared directly to the colour sensitivity of the eye, the luminance signal can be considered as being 'panchromatic', and can be applied to monochrome tubes to give a fair representation of colour scenes. This is a basic element of compatibility.

Fig. 3.5 showed how a monochrome picture tube responds to signals derived from a colour camera. It is assumed here that the grid is biased for beam current

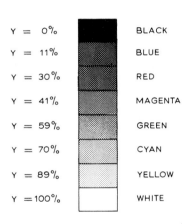

Y	=	0%	BLACK
Y	=	11%	BLUE
Y	=	30%	RED
Y	=	41%	MAGENTA
Y	=	59%	GREEN
Y	=	70%	CYAN
Y	=	89%	YELLOW
Y	=	100%	WHITE

FIG. 3.6. *This scale of greys shows approximate monochrome displays of colour signal derived from a colour camera scanning coloured scenes based on 100 per cent colour saturation.*

cut-off on black (absence of signal), and that the cathode signal swing is negative (as it is in practice). For white, the peak white value of Y is assumed, and the displays are based on this.

Fig. 3.6 gives the same information in a slightly different way. Here the grey scale between black and white corresponds to the primary and complementary colours in order of Y value from blue to red. This scale is useful to keep in mind. Of course, there are many more intermediate greys than can possibly be indicated on such a scale. Remember, the monochrome reproduction of colour signals is just as important in colour television as the display in colours, for it is this compatibility concept that makes the system practicable and an economic proposition.

COLOUR DISPLAY

A device capable of displaying the signals in their colours requires reproduction of the actual colours, both separately and in addition. Much research has been devoted to colour display devices over the years, and many patents have been taken out and have lapsed, but all systems are based on the additive mixing of the three primary hues. The basic system employs three monochrome picture tubes to reproduce three pictures of the same transmission, one seen through a red filter, the second through a green filter and the third through a blue one. When the three images are brought into optical register, the eye sees the picture in full colour.

Fig. 3.7 shows the three tubes arranged to project their images upon a white screen. A practical scheme is similar to that adopted in console rear-projection sets, but multiplied by three. In both cases the optical side must be accurately engineered to ensure that the three projected images fall exactly on top of each other. This is known as *image registration*—a common colour television term. It will be remembered that similar registration is demanded in the colour camera.

When the three tubes are biased together for beam current cut-off on zero signal, the occurrence of video signal will cause the three tubes to display together and, assuming matching characteristics and correct selection of the primary colour filters, the three displays when superimposed will add together and give white and greys.

The effect is perhaps easier to appreciate when plain, unmodulated rasters are considered, rather than changing-level modulation signals. The steady-value signal inputs in Fig. 3.7 imply that each tube is delivering a steady, unmodulated raster, and the three primary colour rasters appear on the screen as a single white raster, when properly superimposed.

Correct registration is achieved when the three rasters fall in exactly the same position, line by line, on the screen. However, registration is not particularly important on plain unmodulated rasters, but it becomes very critical when images are projected, for then the *elements* of each picture have to 'register' exactly on top of each other on the screen. Lack of image registration shows up like a badly printed colour picture when the colours are displaced relative to each other.

We know, of course, that a raster is the subjective effect of a fast-moving scan spot, bringing in the factor of persistence of vision. Monochrome tubes have one scan spot, of course, but a colour display system has three spots compounded (as a 'triad' of red, green and blue dots) to form one spot. If the scanning of the three spots is muted, the display would reveal the three spots or, with correct image registration, one composite spot.

It is not good practice to let any cathode-ray display show bright, stationary spots because this is likely to damage the fluorescent screen, but this is mentioned simply to illustrate the principle and to highlight another point. Spots resulting from muted scanning would superimpose to form one spot in the middle of the screen, but good registration here does not necessarily indicate that the registration is similarly accurate at all other points on the screen. Optical shortcomings and relative linearity and raster deviations can impair the registration severely as the beams are deflected from their mid-screen positions.

This represents a big problem in colour television, not only so far as this simple method of colour display is concerned, but also in the shadowmask colour picture tube.

Good overall registration means, therefore, that the scan spot at any position on the screen looks like one spot but is actually compounded of three spots. It differs from the scan spot on monochrome tubes because it is able to alter in colour in addition to defining picture detail and brightness.

Colours are produced when the red, green and blue signal voltages change in value relative to each other. For instance, if the green signal voltage rises while

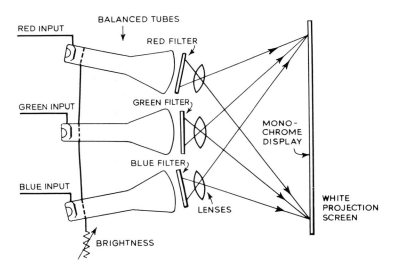

FIG. 3.7. *Simple three-tube colour projection system. White is produced when the inputs are as shown at the tubes and when the luminosity of the screen phosphors are correctly balanced.*

the red and blue signal voltages fall, the display would change from white towards green. In practice, relative colour signal voltage changes are controlled by circuits whose bandwidth is only about 20 per cent of that of the circuits handling the Y signal. Brightness changes thus occur more rapidly than colour changes. Put another way, the brightness or luminance definition is better than the colour definition. This is possible because of the character of the eye in defining objects in colour (Chapter 2).

The whole range of colours in the enclosed colour triangle of Plate 1 can be produced by the arrangement in Fig. 3.7. This, as it is shown, however, would be very inefficient, and something more elaborate is demanded in practice—such as the Philips large-screen colour projection system. This employs three projectors each based on the Schmidt optical system. The tubes used are about 3½ in. diameter and each takes 50 kV final anode potential to give a display sufficiently bright for projection. Colour pictures up to 13 ft by 16 ft can be projected at adequate intensity. The Schmidt system is that used in monochrome projection sets, popular some years back.

Fig. 3.8 reveals the basic elements of a three-tube rear-projection colour system. This uses three of the smaller Schmidt optical systems, as used in the

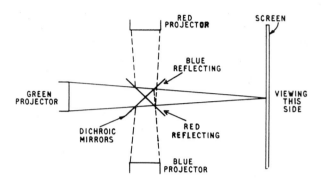

FIG. 3.8. *Showing how dichroic mirrors can be used in a rear-projection colour system.*

monochrome sets just referred to, in conjunction with a pair of dichroic mirrors to pass and reflect light of certain colours, as in colour cameras.

To sum up we have seen that the display in colour—whatever its nature— must create a scan spot capable of changing from black (zero) to peak white, through all the greys, and from one hue to another over almost the whole of the colour spectrum. It should also be possible to deflect the spot in the manner of the single scan spot of monochrome sets. Since the colour-changing property can so far best be obtained by the superimposition of three spots—red, green and blue—there must be some scheme to maintain their coincidence or registration at all positions on the display screen.

Sets employing the shadowmask tube resolve this problem by static and dynamic convergence controls. However, in an endeavour to reduce or avoid the problem, several colour display tubes have been evolved, and taken up to prototype form and beyond, with only a single electron beam and scan spot; but they have not so far succeeded commercially.

The screen of this kind of tube, such as the *Lawrence* (or *Chromatron*) tube, is composed of a series of red, green and blue phosphor stripes arranged in the order of a green stripe between each red and blue stripe. Tubes so far made employing this technique have about 500 green and 250 each of blue and red stripes.

The electron beam is caused to deflect locally, close to the surface of the screen, so as to select the phosphor stripe corresponding to the colour detail at any instant. This local deflection, in addition to the ordinary line and field scanning, is achieved by electrostatically charged wire grids close to the screen.

The charge is arranged to alternate at a specific rate which is tied to the line, field or colour subcarrier frequency, and the colour selection is geared to the decoded colour information by means of an electronic switch.

The switching means that the colour information is displayed one line at a time in swift time sequence. The system is thus called *time-sequential*.

When three electron beams are used in the display device, all the colours are displayed together, at the same time, and a system related to such a display, such as NTSC or PAL, is called *simultaneous*. It is possible to translate simultaneous signals to a sequential display, but it is not very likely that the sequential system will ever take precedence over the existing simultaneous system. It is worth noting that a sequential system was used in America, but was revoked in favour of the simultaneous NTSC system, which has now been in use for many years. However, it is desirable to be aware of the sequential system.

Other display devices using the sequential principle are the *Apple, Zebra* and *Banana* tubes, named after the way they look! These are sometimes referred to

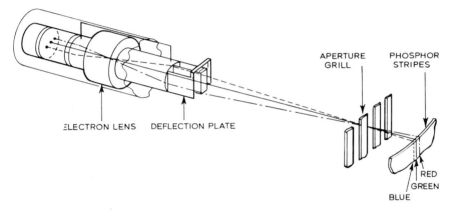

FIG. 3.9. *Principle of the Sony 'Trinitron' single-gun colour picture tube.*

as *index tubes* because of the indexing wires or conductors featured for electronic switching of the colour circuits in synchronism with the scanning of the stripes. The shadowmask tube, however, is the most commonly used at present, but a word about the Banana tube would not be amiss.

In this tube the electron beam is 'wobbled' over the three stripes (all that are used in the tube), and this wobble switches the colour circuits synchronously. The beam is deflected horizontally for the line, while the field scan is obtained by a system of lenses which is caused to rotate axially round the tube. The picture is viewed from a hyperbolic mirror, the three phosphors lighting to give the colours.

SONY 'TRINITRON' TUBE

It is noteworthy that the Japanese Sony Corporation has developed a single-gun colour picture tube called the 'Trinitron' which is in use in some of the

Sony colour receivers. The single gun emits simultaneously three electron beams, but unlike the three-gun shadowmask tube, in which each of the three guns has a small diameter electron lens to focus its beam, the 'Trinitron' arranges for its three beams to be converged and focused through a special electron optical system, which consists of lenses and prisms as shown in Fig. 3.9.

Because the three electron beams are positioned on the same horizontal plane, correct convergence is achieved merely by shifting the two outside beams sideways until they meet the middle beam. The shadowmask tube requires a relatively complex convergence system to obtain similar results owing to the three guns being in delta formation. The 'Trinitron' is converged by means of the electron prisms, and since these deflect the beams electrostatically, dynamic convergence is achieved by a circuit which varies the voltages on the left and right deflection plates of the prisms as the beams themselves are deflected by the scanning coils to provide the rasters.

Colour selection is by the 'aperture grill' shown in Fig. 3.9, which consists of a metal plate formed into vertical strips. These 'guide' the beams on to the screen phosphors which, instead of dots, consist of vertical stripes of alternately blue, green and red phosphors. Good brightness and sharp focus are provided by the relatively large diameter electron lenses and the overall efficiency is said to be higher than the shadowmask owing to reduced electron loss to the 'aperture grill', compared with the loss to the shadowmask. Moreover, the Moiré patterns which sometimes trouble the shadowmask tube due to the array of glowing dots are avoided by the use of the vertical phosphor stripes.

CHAPTER 4

The Shadowmask Colour Tube

THE TUBE USED in domestic colour receivers is the shadowmask type developed in the USA by RCA. Its full name is the three-gun tricolour shadowmask picture tube because it has three electron beams which individually excite three different colour phosphors, each beam being prevented from falling on to the 'wrong' two colours by an inbuilt shadowmask.

BASIC PRINCIPLES

The effective scan spot produced by this tube is really an integration by the eye of three separate spots of red, green and blue. Viewed through a magnifying glass (see Plate 3), the three different colour-glowing spots can easily be seen, but at the normal viewing distance the eye cannot distinguish such tiny points and a mix of the colours is seen.

The screen is composed of a great number of 'triads'—clusters of three—phosphor dots, all of pin-head size. The chemistry of the phosphors is such that, when electrons hit them, one emits red, another green and the third blue light. If a single electron beam was to strike the three phosphors of a triad simultaneously the phosphors would emit their red, blue and green lights together and—assuming their brightnesses were suitably adjusted—the viewer would see a white spot. There would, however, be no way of getting a range of colours.

The shadowmask tube has three electron guns shooting out three electron beams, as shown in Fig. 4.1. The function of the shadowmask, which is a thin steel plate with some 440,000 holes, is to ensure that each beam—after passing through the plate—strikes only the corresponding colour phosphor dots.

As there is a gun emitting a beam for each colour we get such terms as the 'red gun' and 'blue beam'. These, of course, refer to the colour produced at the screen, not to any peculiar property of the gun or beam! Each of the three guns in the tube is similar in action to the gun in a monochrome tube. Thus, it is possible to control the luminance of the effective scan spot by regulating the three beams together, and the hue by regulating them relative to each other.

If the beam currents are first adjusted so that the red, green and blue lights add to give a white scan spot, then the brightness can be varied by adjusting the signal or bias on the three sections equally. The result is then exactly the same as with a monochrome tube. The hue of the spot will change as the bias (or signal) of each gun is altered separately. A red spot will appear with the green and blue guns biased off, a blue spot with the red and green guns biased off, and a green spot with the red and blue guns off. The complementary colours are produced by leaving two guns on and biasing off one—the blue gun off to obtain yellow, the red gun off to obtain cyan and the green gun off to

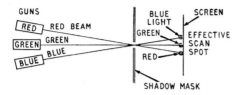

FIG. 4.1. *Electron beams from the three guns strike the correct phosphor dots by passing through holes in the shadowmask.*

obtain magenta. This accounts for white and black, the primaries—red, green and blue—and the complementaries—yellow, cyan and magenta. Intermediate colours are obtained by intermediate biasing of the three guns.

The saturation of any colour depends upon the proportion of white light in the mix. Colours produced in the above manner are fully saturated, because white in any proportion is produced only with the three primaries together, and the above dealt with either one or two guns working.

In most scenes all three guns produce electrons, so colours are rarely, if ever, fully saturated. Fully saturated colours are rare in nature. The sky, for instance, is desaturated blue. On looking at the sky reproduced on a colour tube through a magnifying glass it will be seen that all three phosphors of a triad are glowing, but with blue predominating. The result is a desaturated blue. Likewise with pale yellow. All three phosphors will be glowing, but yellow will predominate and this will yield a desaturated yellow.

Having seen how the scan spot of a shadowmask tube can register luminance, hue and saturation, let us see the basic method of control used on the tube for these actions.

One way of connection is for the Y signal to be applied to the cathodes and the red, green and blue colour-difference signals (page 49) to be applied separately to the corresponding grids, as shown in Fig. 4.2. The Y connection to the cathodes is analogous to the video connection in monochrome tubes between the video amplifier and the tube cathode. The luminance amplifier is then the counterpart of the video amplifier in a black-and-white set.

The colour-difference signals are derived from separate red, green and blue colour-difference amplifiers and d.c.-coupled to the grids, which means that the grids are at some positive potential. This is indicated in Fig. 4.2 by resistors R_V being connected to a positive source. Likewise, the cathodes are d.c.-coupled to the luminance amplifier, which means that the cathodes are also at some positive potential.

The d.c. potentials at grids and cathodes are arranged so that the brightness control at minimum brightness setting makes the grids considerably negative with respect to the cathodes and so cuts off the beam current. At 'higher' settings the brightness control restores the beam current by making the grids less negative relative to the cathodes.

Since the cathodes are d.c.-coupled to the anode of the luminance amplifier, a change of signal in this amplifier will alter the potential at its anode and at the cathodes of the shadowmask tube and so effect a change in the 'biasing' of the guns—all three together.

How do these grid and cathode signals interact? And why, in fact, are signals put on grids *and* cathodes? The answers come from the nature of the colour-difference signals (see page 49). These are the 'real' or original primary colour

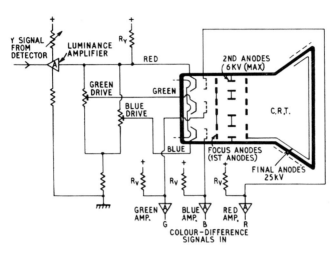

FIG. 4.2. *Y signal is applied to the tube cathodes together and the colour-difference signals separately to the red, green and blue grids. Preset adjustments are provided to obtain a 'white' display on Y signal only.*

signals of the TV camera with the Y, the luminance signal, subtracted. The colour-difference signals exist up to the picture tube grids, and this gives the problem of how to get back to the primary colour signals—the red, green and blue signals corresponding to those given by the camera.

What is needed is simply the recombination of the Y signal. For instance, if we take $R - Y$ and subtract $-Y$ from it we get back to R—the primary colour. Some colour receivers use a matrix circuit for each colour-difference signal for this purpose but the picture tube itself is often left to do the job.

CONSTANT LUMINANCE

With colour scenes, the colour-difference signals provide for both hue and saturation; the Y signal for the brightness (luminance). In the limiting case of a black-and-white scene there will be no colour-difference signals and the Y signal on the tube cathodes will produce the monochrome picture. This will be monochrome, of course, because the three guns are initially biased to produce red, green and blue lights in the proportions to give white light. This is the basis of the *constant luminance* colour television principle; that is, a true chroma signal is one which defines the hue and saturation of the picture only, with the Y signal alone being concerned with the luminance or brightness of the picture. In practice there is some small departure from this ideal constant luminance condition.

Adjustments are provided to counter tolerance variations in valves or transistors, other components and in the picture tube itself, so that a white light is produced by a 'white' signal. It is also necessary to adjust the beams to match the colour phosphor 'sensitivity' in terms of white light output. The circuit in Fig. 4.2 provides for such trimming by the green and blue drive presets. Full signal is applied to the cathode of the red gun. These presets are initially adjusted to give a white raster over the luminance range.

We have studied the shadowmask tube so far purely as a device for producing a scanning spot of variable luminance (brightness), hue and saturation. Changes in the relative levels of the colour signals applied to each gun thus swing the hue of the compounded scanning spot over the colour triangle, while changes in Y input control the luminance, with colour saturation being controlled by the presence of 'white signal'—not one or two but *three* beams contributing in some proportion to the light emission.

Now let us briefly look at the tube under conditions of beam deflection (detailed information is given later). There are nearly half a million phosphor

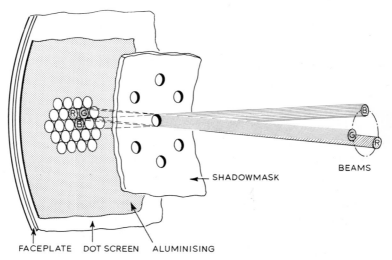

FACEPLATE DOT SCREEN ALUMINISING

Fig. 4.3. *The action of the shadowmask. Note that the three beams pass through one hole and then diverge a little between the shadowmask and the screen, so that each strikes its corresponding colour phosphor.*

triads on the inner surface of a shadowmask screen (over a million phosphor dots in all). The guns are set at a small angle to each other, and the three beams aimed so that they converge at the shadowmask, shown in Fig. 4.3. The beams then diverge a little between shadowmask and screen, each beam striking its corresponding colour phosphor of a triad. The angle of the guns to each other produces the required divergence of the beams relative to the distance between the shadowmask and screen and to the displacement of the phosphor dots in a triad. The great need for extreme accuracy can be appreciated.

Not all the beam electrons pass through a hole. The diameter of the beams is reduced, thereby ensuring that bombardment is restricted to those phosphors of a triad seen, as it were, by the beam through the corresponding hole in the shadowmask. This makes it easier to avoid colour fringing.

Since the electron 'pencils' are reduced to a small diameter by the shadowmask holes, more power is required in the beams for a given light output. This is one reason why the final anode potential of a shadowmask tube is of the order of 25 kV.

The shadowmask assists in providing the correct registration of the beams and their corresponding phosphors as they are deflected vertically and horizontally to produce the raster. Magnetic deflection is used, as with monochrome tubes, but a greater scanning power is needed owing to the greater 'stiffness' of the beams caused by the higher e.h.t. (and the consequent greater electron acceleration).

NEED FOR CONVERGENCE

For the three colours to register accurately at all points on the screen, the three beams must always converge exactly at the shadowmask. Centre convergence is adjusted by the fields of a permanent magnet system on the tube neck. This sets the effective angles of the beams a little, relative to each other, and is called *static convergence*.

Because the radius of the screen and the shadowmask differs from the radius over which the three beams are deflected, from the deflection centre within the fields of the scanning coils on the tube neck, progressive 'mis-convergence' occurs as the beams swing away from the centre of the screen. This is corrected by changing magnetic fields which are derived from the line and field timebase currents and delivered to a dynamic convergence unit, also located on the tube neck. Both *dynamic convergence*, as this is called, and static convergence are examined in greater detail later.

Colour-tube phosphors are of relatively short persistence, meaning that they retain their illumination for only a short time after the exciting electron beams have passed on. Long persistence would cause streaks of colour following the edges of a moving image—a shortcoming of early shadowmask tubes.

Highly efficient sulphide phosphors are now employed and these give brighter colour pictures than hitherto obtainable. A major problem was to obtain sufficient illumination from the red phosphor. A very high red beam current was needed to match the illumination from the green and blue phosphors to give white. This difficulty has been eased by the development of a rare-earth red phosphor. Subsequent improvements have also been made in the relative efficiencies of green and blue phosphors.

DOT SCREEN AND SHADOWMASK CONSTRUCTION

The phosphor dots are deposited on the screen in the sequence green, blue and red before the cone is welded to the screen. The shadowmask is used as a negative for the positioning of the dots. For instance, the green dots are obtained by the application of a uniform layer of the green phosphor material to the screen along with a photosensitive resist dried by infra-red heat. The shadowmask—acting as a negative—is positioned carefully over this coating and the coating exposed to ultra-violet radiation.

This bombards the coating through the holes in the shadowmask at exactly the same angles as will be taken by the green beam electrons themselves when the tube is working, resulting in pin-point ultra-violet exposures 'locking' the coating at these points. The remaining coating is then washed off, leaving only the green phosphor dots. The process is repeated for the blue and red phosphors.

The shadowmask is of sheet steel and the holes are made by an etching process. The sheet is given a coating of light-sensitive material and exposed to

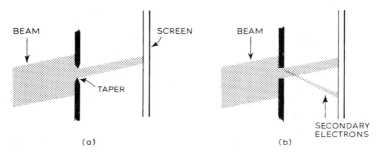

FIG. 4.4. *Tapered holes (a) avoid secondary electrons (b).*

light through a negative master dot pattern. This produces a resist, except at the holes, so that when the sheet is exposed to acid the holes are etched out.

The holes are tapered as shown at (*a*) in Fig. 4.4 to prevent a spray of secondary electrons developing when the beam strikes the edges. The spray that would occur in a non-tapered hole is shown at (*b*). The effect of this would be a slight 'ghosting' of the wrong colours round points of sharp detail.

COLOUR TUBE PARAMETERS

Colour tubes are commonly 49 cm (19 in.), 55 cm (22 in.) and 63 cm (25 in.), corresponding to the maximum screen diagonal. The aspect ratio (width-to-height) is the conventional 4:3 approximation (a 55 cm tube, for instance, having a width of 447 mm and a height of 337 mm), giving a match to 35 mm film frames. Some of these tubes are called 'supersquare' because the faceplates are flatter and more rectangular than those of earlier tubes with an aspect ratio of about 5:4. Most of them are now designed for push-through cabinet presentation. The weight of a 55 cm tube is about 15 kg and neck diameter 36·5 mm.

An inactive tube appears to have the smooth off-white screen of a monochrome tube. Close scrutiny through a magnifying glass is needed to reveal the multiplicity of very small phosphor spots. They almost touch each other, and when passive they all look the same colour in spite of their ability to emit red, green and blue light.

The potential of the final anode is close to 25 kV and so x-ray emission is a possibility, as it is with the small tubes of monochrome projection sets running at about the same potential. However, the screen of the shadowmask does not project beyond the cabinet, but is arranged to be flush with the front (as with contemporary 'push-through' black-and-white tubes) and the screen itself is made of special glass (cerium oxide) to absorb x-rays. Ordinary glass would be darkened by the x-rays, but this glass remains clear throughout its working life.

The phosphor-dot screen is rear-aluminised, as are monochrome screens, to prevent ion burn and to obviate the need for electrostatic or magnetic ion-trap gun assemblies. These would, indeed, complicate matters with three guns whose beams have to be very carefully controlled.

A colour tube is secured to the cabinet by four mounting lugs at the corners of a strong metal band round its rim. The band is also for 'implosion-proofing' as with the latest monochrome tubes.

Band and mounting lugs are clearly shown in Fig. 4.5, a picture of Mullard's 63 cm tube. To the right of this picture are important '500-series' colour valves by Mullard. They are the PL505 line output pentode, PD500 shunt stabiliser triode, PY500 booster diode and GY501 e.h.t. rectifier.

An opened up Mazda shadowmask tube (Fig. 4.6) shows the neck flare, shadowmask and screen. The shadowmask, a sheet of thin steel, is located a little over 1 cm from the phosphor-dot screen.

Expansion and contraction arising from temperature change are catered for by the mechanics of the mask's mounting, for it is most important that the holes in the mask always remain in accurate alignment with the phosphor triads. Even slight displacement with respect to the phosphor dots would impair colour registration. Displacement and buckling are avoided by the mounting moving the mask very small distances along the tube axis when there are dimensional changes.

The shadowmask certainly warms up when the tube is running, so there is no doubt about expansion. This arises from the energy liberated when the 'surplus' electrons of the three beams hit the mask. About a quarter of the electrons of each beam pass through the holes in the mask; those remaining heat the steel.

TUBE POTENTIALS

Each beam may carry a current of about 320 μA, with about 80 μA contributing to light emission. The remaining 240 μA of beam current (at 25 kV) applies some-

FIG. 4.5. *Mullard 'Panorama' 90 degree colour tube, showing corner fixing lugs. The valves are Mullard 500-series, PL505 line output, PD500 shunt stabiliser triode, PY500 booster diode and GY501 EHT rectifier (courtesy, Mullard Ltd.).*

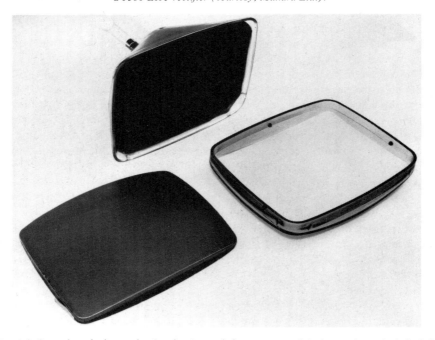

FIG. 4.6. *Opened-up shadowmask tube, showing neck flare, screen and shadowmask, in which the holes are too small to show up (courtesy, Mazda Colour TV Tubes Ltd.).*

thing like 6·2 W to the mask—or 18·6 W with the three beams running at maximum current. Total light emission is limited to three times 80 µA, or 6·2 W (at 25 kV), which is about the same as the loss in *each* beam to the mask.

The three electron beams 'cover' about three holes, thereby energising three corresponding tri-phosphor clusters. The scan spot is really an integration of red, green and blue lights from these three triads.

The tubes have about 90 degrees angle of deflection* (the Mullard 22 in tube has 92 degrees deflection); this is the total beam deflection from one corner of the screen to the corner diagonally opposite. The three beams are deflected in unison magnetically by means of scanning coils on the tube neck.

Besides the scan currents, d.c. is passed through the line and field coils to provide horizontal and vertical picture shift. Preset shift controls regulate these currents. A similar scheme used in early monochrome sets was superseded by permanent-magnet shift assemblies. These cannot be used on colour tubes, however, for fear of 'misconverging' the three electron beams. Nevertheless,

FIG. 4.7. *Deflector coil assembly for 90 degree colour tubes. Note the wing-nut adjustment (courtesy, Mullard Ltd.).*

there are similar magnets on the colour tube for purity and convergence adjustments, discussed later. A scanning assembly for colour tubes is shown in Fig. 4.7.

The base of the Mazda shadowmask tubes (CTA2550 63 cm and CTA1950 49 cm) is seen in Fig. 4.8(*a*). All fourteen pins, excepting pins 8 and 10, are used. Pin 9, carrying low e.h.t. (about 5 kV) for anode 2, thus has less chance to flash across to adjacent 'live' pins. Fig. 4.8(*b*) is the colour tube symbol together with typical electrode operating potentials (for each gun). The first anode (a_1) is for initial beam acceleration and is arranged for separate adjustment on each gun (a_1R, a_1G and a_1B).

The potential here influences the grid voltage (V_g) for beam current cut-off (raster cut-off). With V_{a1} at 300 V, beam current cut-off occurs with the grid

* Tubes of wider scanning angle are now being used, see pages 318, 319 and 333.

between -70 V and -140 V. Thus preset controls for each a_1 electrode supply allow the gun sensitivities to be balanced.

The second anodes (a_2) of the three guns are internally connected and the common connection brought out to pin 9. This is the focusing electrode and a preset potentiometer swings its potential over the range $4\cdot2$ to 5 kV—considerably higher than the focusing potential of a monochrome tube!

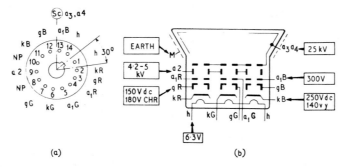

(a)   (b)

FIG. 4.8. *Base connections of the Mazda colour tubes (a) and colour tube symbol (b), showing typical operating voltages.*

Third and fourth anodes (a_3 and a_4) comprise the final anode system taking full e.h.t. These are also connected together internally and the common connection is brought out to the e.h.t. connector on the flare. E.h.t. is in the range 20–27·5 kV. The third anodes are part of the gun assemblies while the fourth anode is an internal conductive coating on the flare. This is in connection with both the aluminising behind the phosphor dot screen and the shadowmask as

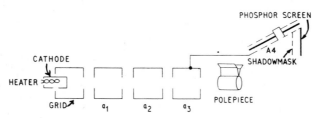

FIG. 4.9. *25 kV potential is applied to the final anode comprising a_3 and a_4 together. This anode is also connected to the internal conductive coating (forming part of the e.h.t. reservoir capacitor) and to the aluminising behind the phosphor dot screen.*

shown in Fig. 4.9. This also shows the approximate disposition of the electrodes of each gun assembly.

A picture of the Mazda tri-gun assembly is given in Fig. 4.10. The three heaters are rated at 6·3 V and (0·8 A) for the 49 cm tubes and 6·3 V (1·3 A) for the 63 cm tubes. These currents mean that the three heaters are wired in parallel, not in series as implied by the symbol.

We have seen that the three cathodes are together d.c. connected to the Y amplifier. This brings them to about 250 V d.c., upon which is superimposed the Y signal with a maximum peak to peak swing of about 140 V.

The three grids are separately d.c. coupled to the corresponding colour-difference amplifiers. Each grid is brought to about 150 V d.c. average (depending on preset adjustments) with maximum 180 V peak to peak signal swing of chrominance information. Thus, under zero signal conditions, the effective bias between each grid and cathode is about -100 V, which is in the range for raster cut-off. Final 'tailoring' is achieved by main and preset brightness controls.

The $-Y$ and 'colouring' signals together reduce the bias below the cut-off value, and the three guns fire electrons at their corresponding phosphors to create red, green and blue lights to match the luminance and colour information in the signal.

Electrode capacitances are important. For the 49 cm and 63 cm models they are typically: grid–all 7 pF each gun, cathode–all 11 pF, a_2–all 5 pF and $a_3 + a_4$– external conductive coating 2000 pF. The latter constitutes the e.h.t. smoothing

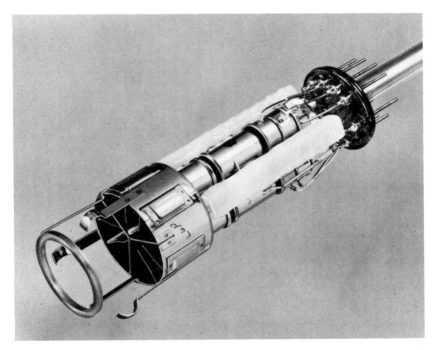

FIG. 4.10. *Three-gun assembly of Thorn colour tube (courtesy, Mazda Colour TV Tubes Ltd.).*

(reservoir) capacitance with the glass of the flare acting as the dielectric and the inner and outer conductive coatings as the two plates. The internal coating is connected to e.h.t. as we have seen, while the outer one is 'earthed'.

Absolute tube ratings include: peak heater/cathode potential during warming-up period not exceeding 15 s, 410 V; running heater/cathode potential (heater negative), 250 V; maximum grid/cathode resistance, 750 kΩ; and maximum a_2

resistance, 7·5 MΩ. Dimensions of the Thorn-A.E.I. 63 cm tube are illustrated in Fig. 4.11 ((d) shows all the accessories required on the tube neck—detailed later).

The screen is grey-coloured glass to keep the picture contrast up in extra bright viewing environments. Light transmission is about 50 per cent and the

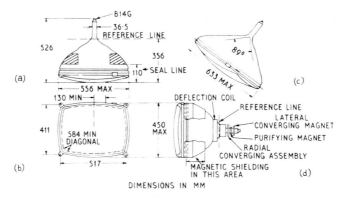

FIG. 4.11. *Dimensions of Thorn 63 cm (25 in.) colour tube. Diagram (d) includes location of scanning coils and convergence components on the neck.*

light output at the middle of the screen with the three beams together running at about 800 μA yields around 71 nt, the luminance becoming greater with new designs (nt is the symbol for 'nit', metric units of luminance).

CONVERGENCE ASSEMBLIES

Electrons issuing from each gun pass through a 'radial convergence' pole-piece made of metal affected by magnetism. The pole-pieces are within the tube neck and their purpose is to concentrate the magnetic fields—applied from the radial convergence assembly outside the tube neck—around the three beams to provide them with components of corrective deflection. This is called convergence and will be considered fully in Chapter 6. In the meantime a brief description will enable the shadowmask tube to be more fully understood.

A sectional view of the pole-pieces of all three guns is seen in Fig. 4.12(a), and (b) shows the positions of the guns relative to the tube base. Diagram (c) shows how the poles of the radial convergence assembly are aligned to the internal pole-pieces for optimum magnetic coupling through the glass. The poles can also be seen in Fig. 4.10.

The internal fields are developed as indicated by the broken lines in Fig. 4.12(c) and they subject each beam to a deflecting force which causes a radial movement (shown by the arrowed lines). The beams are influenced only by their appropriate magnetic fields, interaction being avoided by the nature of the internal pole-pieces and the shielding.

Convergence magnetic fields are provided both by permanent magnets, whose effective transferred field can be adjusted by magnet rotation or some other means, and by field coils energised by current waveforms. Fig. 4.12(c) also shows these field sources, and a complete radial convergence assembly (Mullard) is shown in Fig. 4.13. The three adjustable permanent magnets can be seen at the extremes of the three limbs. The field coils themselves are hidden beneath the plastic mouldings. The windings are terminated at the wire lead-outs along one side of each limb.

The three adjustable fields from the permanent magnets correct the line of travel of the three beams so that they all converge to a single point near the

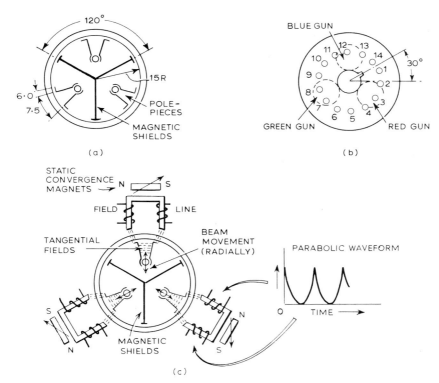

Fig. 4.12. *Sectional view of gun pole-pieces (a), location of guns relative to base of tube (b) and nature of fields produced by dynamic convergence coils and static convergence magnets (c), showing also the current wave in the coils.*

middle of the screen. Without such correction the three beams would fail to pass through the correct holes in the shadowmask and three displaced red, green and blue scan spots would be displayed instead of an integrated 'single' spot.

NEED FOR BLUE LATERAL CONVERGENCE

The three fields of the permanent magnets cannot alone give perfect middle-of-screen convergence, because each field can only deflect its appropriate beam along a straight line. While it would be possible to get two scan spots (say, the red and green) to register at the same point near the middle of the screen, this point of coincidence might not be on the path along which the third (blue) scan spot would move when the third magnet was adjusted. However hard one tried, the third spot would then never register exactly with the other two.

This problem is resolved by the use of a separate magnetic assembly, containing a permanent magnet and sometimes an electromagnet also. This allows

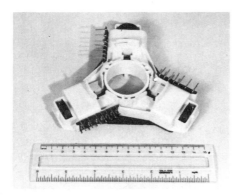

FIG. 4.13. *Mullard convergence unit for 90 degree, 63 cm (25 in.) colour tube (courtesy, Mullard Ltd.).*

one beam to be moved in a second direction, at an angle to the movement provided by the appropriate field of the radial convergence assembly. The blue beam is selected for this two-way movement, and the device giving the second movement is called the blue lateral convergence assembly, or simply the lateral convergence magnet.

An assembly of this kind is shown in Fig. 4.14. The single adjustable permanent magnet at the top gives the blue beam a lateral bias as required for correct

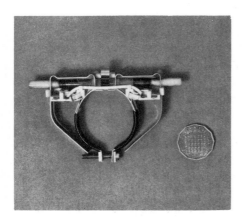

FIG. 4.14. *Blue lateral convergence assembly (courtesy, Mullard Ltd.).*

mid-screen convergence of the blue scan spot. When the permanent magnet is adjusted, the strength of the field, as seen by the blue beam, alters and the beam deflection can then be regulated. Field windings are enclosed in plastic on either side of the permanent magnet, and the assembly as a whole slips over the tube neck something like the ion trap magnet of early monochrome tubes.

It is worth remembering that an electron beam is virtually a flow of current through space and, as with a conductor passing current in a magnetic field, the deflecting force is always at right-angles to the magnetic field. The radial convergence fields are tangential to the tube axis and thus cause the beams to deflect into or away from the tube axis when the magnets are adjusted. The blue lateral field is such that the beam is deflected *across* the tube axis.

PRINCIPLES OF CONVERGENCE ADJUSTMENTS

The first step in static convergence adjustment is to get the red and green beams to converge in the middle of the screen by adjusting the permanent magnets of the radial convergence assembly and then to get the blue beam to register with these two by adjusting the blue radial and lateral magnets (see page 73). Since it is not possible to mute both timebases to secure a spot display at screen centre, a colour pattern generator, providing grid pattern and sometimes dot signals, is used to produce a suitable signal for making the adjustments.

The permanent magnets provide only convergence in the middle of the screen, called *static convergence*. Convergence from the centre outward, however, is called *dynamic convergence*. All convergence adjustments demand a crosshatch and dot generator—or a test card transmission—and even then they can be tricky. Ways and means of making these adjustments are more fully described in Chapter 6.

As the timebases cause the three beams to be deflected (in unison) away from the middle of the screen, the tendency for the three to part company is aggravated, even when the static convergence has been accurately carried out. This is partly because the angle of radius of the screen differs from the beam deflection angle, and partly because the beams 'hinge' from different points due to the different gun positions. The tendency for the beams to separate from each other is worse towards the extreme edges. Something more sophisticated than fixed magnetic fields is required to hold the three 'scan spots' coincident over the whole of the shadowmask.

This is where the coils on the convergence assembly take effect. Each limb of the radial assembly carries a pair of coils. One of each pair is fed with modified line scanning current and the other of each pair with modified field scanning current (Fig. 4.15). The winding of the lateral assembly is fed only with modified line scanning current.

The modified scanning currents are derived from a convergence circuit which itself is energised by the timebases. The resulting waveforms are essentially parabolic. The magnetic fields so produced change the magnetic convergence in synchronism with the scanning so that the progressive geometric unbalance of the three beams is corrected by changing (dynamic) magnetic fields as the beams sweep the screen.

No dynamic correction is required when the beams are central (i.e. zero deflection), since the static system has ensured convergence, but progressively more correction is needed as the beams sweep wider.

This is why parabolic current waveforms are used (Fig. 4.15). They give a greater deflection current component at start and finish, relative to centre.

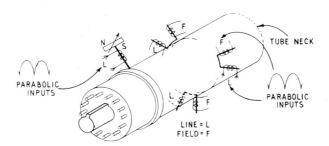

FIG. 4.15. *Showing the positions of the convergence assemblies round the tube neck. Small diagrams show waveforms applied to them for dynamic convergence.*

Some modification, even to this, is required because the three guns do not lie on a common axis. The beam from the top gun (blue), for example, must sweep over a greater distance downwards than the beams from the other guns. Similarly, the red beam must sweep more to the left than to the right of the screen, and conversely with the green beam.

Fine adjustments to the current wave shapes are made by preset controls in the convergence circuit. Adjusting these controls properly represents one of the biggest skills needed in the setting-up of colour sets, especially dual-standard models.

PRINCIPLE OF PURITY

The only neck accessory we have not yet looked at is the purity magnet. The term 'purity' describes the ability of the tube to produce a raster of pure colour when only one beam is running. The raster is said to be 'pure' when it is, for example, perfectly red with the blue and green beams biased off. If the red beam hits some of the blue and green phosphors, however, the tube is said to be 'impure'.

A tube is made pure by adjusting the purity magnets. These are ring magnets which are similar to those used for picture shift on monochrome necks. There are usually two and they are adjusted by rotating them together or relative to each other. They give a magnetic field across their diameters, and hence through the tube neck, cutting the three beams.

Fig. 4.16 shows at (*a*) how the three beams can be effectively rotated together by turning the two rings together and at (*b*) how the field strength can be adjusted by rotating the rings relative to each other, thereby causing the beams to move at right-angles to the field over a distance depending upon field strength and a direction depending upon field polarity.

The purity magnets affect the three beams together (like the scanning fields) and the idea is to get the beams to emanate from a centre along the neck where

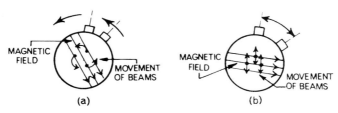

FIG. 4.16. *Purity magnets. Turning both together rotates the three beams simultaneously (a), and rotating them relative to each other (b) varies the field strength and causes the beams to move at right-angles to the lines of force.*

they can then commence a course which will ensure the lighting of their appropriate phosphors on the screen. The setting of the scanning coils on the neck can influence the purity to some extent, but mainly towards the edges of the screen.

NEED FOR DEGAUSSING

Stray and external magnetic fields can badly affect a shadowmask tube and for protection the tube is fitted with a magnetic shield round its flare. The metal rim band protects the screen. The shields, however, can themselves become magnetised. Even the weak field induced into the shields (or other steel members inside the set near the tube) from the Earth's magnetic field can put the tube out of adjustment. Small fields have no practical effect on the single beam of monochrome tubes, but they affect the three beams of a colour tube differently and give convergence and purity troubles; and the effects can alter as the set is moved.

To reduce the effects of extraneous fields and change of orientation, the tube is commonly fitted with the blue beam uppermost and facilities are provided for deleting all signs of residual magnetism in the shield and associated metal members. This is called *degaussing*.

On early sets degaussing was carried out by a large coil of wire connected across the mains supply, called a degaussing coil. This was placed near the tube and metal members, switched on and then gradually moved away and, when out of range, switched off. The effect is the same as demagnetising the head of a tape recorder with a demagnetiser. The changing magnetic field pulls out all the residual magnetism by taking the metal through stages of plus and minus saturation; to avoid switch-off current from putting back the magnetism, the device is not switched off until out of magnetic range.

The latest idea is to pass a mains current through a degaussing coil which is actually part of the set and often placed round the tube flare. When the set

45

is switched on, or sometimes when switched to a channel of different line standard, the current builds up quickly but then drops away slowly. The residual magnetism thus gets less and less and is zero when the current stops. Automatic degaussing ensures that the tube is treated each time the set is used. It is often necessary still to use an external degaussing coil when a colour set is first installed or when moved from one position to another, even though automatic inbuilt degaussing is featured. However, a *very* strong external degaussing field can, in some instances, tend to distort the shadowmask, and only degaussers yielding a moderate field should be used.

CHAPTER 5

An Overall View of the Colour System

THIS CHAPTER is intended to give the reader a bird's-eye picture of the colour television system as a whole, from transmitter to receiver. Block diagrams are used to show how the various sections fit together, and each section is high-lighted in the chapters that follow in terms of circuits, faults and adjustments. This chapter, then, represents an overall plan of the system, revealing the main stages, the way that they work and the signals in them.

One implication of colour television compatibility is that the vision signal must contain the usual monochrome information, blanking and sync pulses required by black-and-white sets. The monochrome signal is, in fact, so little altered that the circuits of a colour set are virtually identical to those of a black-and-white set right up to the vision detector stage.

Sandwiched in with the monochrome signal is the chrominance (chroma for short) information from which colour sets work, and this is sorted out beyond the vision detector stage. It is from the chroma information that the video colour-difference signals are derived.

It will become apparent later that the processing is such as to yield three colour-difference signals but only two of them are actually transmitted via the chroma signal. The third one is recovered at the receiver by adding the other two in a certain manner. After passing through separate luminance and colour-difference circuits, the signals are brought together again at the shadowmask tube.

The scanning and power supply circuits of colour sets are basically the same as those of black-and-white sets, but more powerful and with e.h.t. stabilisation. The line and field timebases are also stabilised to ensure that the rasters remain constant should the power load or supply voltage vary. The timebases also work dynamic convergence circuits.

The starting point of the colour and luminance signals is at the camera which, as was explained in Chapter 3, has three or four camera tubes instead of the usual one in monochrome cameras. Each camera tube scans exactly the same image, and this is converted into video signals in the ordinary manner by beam scanning at 625 lines (50 Hz field in the U.K.). Optical filters are interposed so that only the red rays of the light from the scene activate one tube, green rays the other and blue rays the third.

The three outputs—now referred to as red, green and blue (R, G and B) primary colour signals—are used separately for colour and together in certain proportions for luminance. Unless the scans of the three or four tubes are identical, addition of the R, G and B signals (superimposing three pictures, as it were) degrades the definition. The camera with four tubes has one solely for the luminance signal—giving the monochrome signal for black-and-white sets.

Chapter 2 has revealed that the eye has the ability to distinguish fine details in a scene lit by white light far better than when coloured lighting is used. This means that the eye fails to discern fine detail in one colour alone; it needs 'monochrome' lighting in which all colours are present. In view of this, a satisfactory amount of colour information for colour television can be carried in a band extending only to about 20 per cent of that required by the high-definition luminance channel.

In other words, colour television works by virtue of the full detail of a picture being transmitted in the wideband luminance channel (equal to the ordinary vision channel in 625-line black-and-white sets), while the colour is 'brushed in boldly' at reduced bandwidth.

Indeed, it is because the eye is so easily deceived that monochrome and colour reproduction is possible by the controlled display of red, green and blue dot-triads on the screen of the shadomask tube. Chapter 4 explained how when all the dots are lit in correct intensity white light is seen, while relative changes in intensity produce coloured light.

A colour television system could be devised by feeding the red, green and blue primary colour signals from the camera to separate transmitters, each on a different channel, then employing three-section receivers to process the signals separately. In this way the original red, green and blue signals could be recovered at the set and fed to the grids of the shadowmask tube to give a colour picture.

This is analogous to the early stereo-by-radio idea where two separate transmitters were used, one for the left channel and the other for the right one. Too much radio space was taken up by this scheme and the problem was solved by the present multiplexing technique, where the left channel information and the right channel information are modulated together on one v.h.f. carrier wave in a way that facilitates reproduction of the two channels in isolation at the receiver.

The colour information of television is similarly processed, and technicians who are already conversant with stereo multiplexing should have no undue difficulty in grasping this part of colour television.

A three-channel system, even if it were feasible, would not satisfy the requirements for compatibility, and because only a portion of the total bandwidth is needed for colour it can be contained within the monochrome bandwidth, thereby giving one aspect of compatibility.

COLOUR TRANSMITTER

Fig. 5.1 gives a simplified block diagram of a colour transmitter. The three primary colour signal voltages from the camera (E_R, E_G and E_B) are first fed to a circuit which corrects for *gamma*. This compensates for the non-linearity of the display tube. Without such correction the picture highlights would be stretched and the low-key levels would be compressed, thereby yielding a non-linear contrast ratio.

The corrected signals are given the symbols E_R', E_G' and E_B'—the primes indicating that gamma correction has been applied. The signals are next

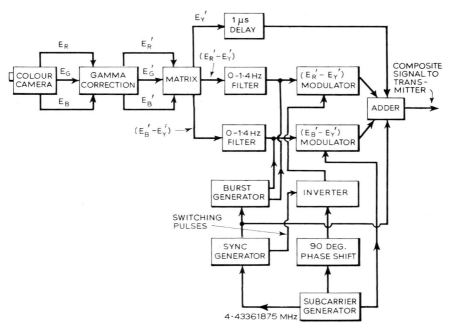

Fig. 5.1. *Block diagram of basic PAL transmitter.*

applied to a matrix which has two functions. First, it adds correctly proportioned signals to produce the gamma-corrected Y signal. Secondly, it eliminates the green signal, leaving the red and blue ones only for transmission. It is fairly easy to put two signals carrying colour information on one subcarrier, but not three. The green signal can be recreated easily from the red and blue signals at the set, as shown later.

COLOUR-DIFFERENCE SIGNALS

The scheme used by both NTSC and PAL converts the primary colour signals to colour-difference signals. These are not signals representing differences between the primary colour signals themselves, but are the differences of two of the primary colour signals (red and blue) each with respect to the Y signal. They are known by the symbols $E_R' - E_Y'$ and $E_B' - E_Y'$ and also respectively as the red and blue colour-difference signals. The green colour-difference signal $E_G' - E_Y'$ is not transmitted, as we have seen, but is recovered at the set.

To make the text easier to read, the letters R, G, B and Y will be used from now on to represent the gamma-corrected E_R', E_G', E_B' and E_Y' signals, unless indicated to the contrary. This will also make formulae appear less complex, but readers should remember that the unprimed letters refer to gamma-corrected signals.

DERIVING THE Y SIGNAL

We have seen that the Y signal is a proportioned sum of the red, green and blue primary colour signals, but at the *output of the camera* they are of equal level, adjusted on white to a reference standard, often taken as 1 V. However, and this is important, when the R, G and B signals are 'added' in the matrix to yield the Y signal, they are added in the proportions red 30 per cent, green 59 per cent and blue 11 per cent of the reference standard. Thus, when the reference is 1 V, red is 0·3 V, green 0·59 V and blue 0·11 V, and when they are added they give Y a value of 1 V, too.

These proportions were originally derived on the basis of the signals produced by a monochrome camera televising in turn fully saturated red, green and blue screens of 'standard' illumination. This means that a monochrome camera produces the same signal voltage ratios from colour scenes as given by the Y output of the transmitter's matrix when a colour camera is televising the same colour. The Y signal thus contains all the monochrome information of the scene, whether it be in colour or black-and-white. This is the luminance signal which occupies the whole of the system's vision bandwidth. It contains all the information on the detail and brightness of the picture. This is why it is possible to turn off the colour at the set, leaving a picture in monochrome which is perfectly viewable and contains virtually all the detail. If the luminance is turned off or fails, however, the resulting display in colour only is of very low definition.

The Y signal makes it possible for a black-and-white set to reproduce a colour transmission in monochrome. There is a very small monochrome 'interference effect', but this does not detract from the entertainment value of monochrome viewing from a colour transmission.

It was explained in Chapter 3 how the value of Y alters with change in output from one or more of the camera tubes; it was also shown how Y can have the same value with many individual values of R, G and B primary-colour signals. The latter means that the same shade of grey will appear on monochrome sets for a range of hues on a colour set.

Now let us return to the block diagram in Fig. 5.1. The Y signal from the matrix is fed through a delay line to the r.f. stages of the transmitter and then to an adder. The colour-difference signals also eventually arrive at the adder, but they are first submodulated, as we shall see later.

First, the question arises as to why the primary-colour signals are converted to colour-difference signals and not modulated direct. There are several aspects of this, all closely related to the development of the present-day colour system. For one thing, colour-difference signals arise only when there is colour in the televised scene. If a colour camera is scanning a scene devoid of colour, then there is only a Y signal. This can be understood by considering the derivations of the Y and red and blue colour-difference signals.

$$Y = 0{\cdot}3R + 0{\cdot}59G + 0{\cdot}11B \qquad \ldots (1)$$

$$R - Y = 1{\cdot}00R - (0{\cdot}3R + 0{\cdot}59G + 0{\cdot}11B)$$
$$= 0{\cdot}7R - 0{\cdot}59G - 0{\cdot}11B \qquad \ldots (2)$$

$$B - Y = 1{\cdot}00B - (0{\cdot}3R + 0{\cdot}59G + 0{\cdot}11B)$$
$$= -0{\cdot}3R - 0{\cdot}59G + 0{\cdot}89B \qquad \ldots (3)$$

The Y signal derivation as we know it is given at (1), the red primary colour signal from which Y is subtracted is given at (2), and the blue primary colour signal from which Y is subtracted is given at (3). Note that the actual primary colour signal by itself is referred to unity (i.e. 1·00). This is because the signals in Y are proportions of unity.

Now, because monochrome inputs to the colour camera give R, G and B primary colour signals in the proportions shown in (1), these *add to zero* when referred to red and blue colour-difference signals, as shown in (2) and (3).

This is an important aspect of the colour system, for it means that colour signals are present in the system only when there is colour in the scene.

CHROMINANCE SIGNAL

The R − Y and B − Y signals are modulated simultaneously in a special way on to a carrier wave which is placed in the video spectrum. This is called the *subcarrier*, but as such it does not itself actually modulate the main carrier wave of the transmitter. What happens is that the process of modulation deletes the subcarrier leaving only the effective sidebands of the R − Y and B − Y signals. These sidebands modulate the main carrier wave along with the ordinary video information. This so-called *suppressed carrier* modulation is also used in stereo radio broadcasting.

The R − Y and B − Y signals quadrature modulate the subcarrier, a function which is considered in detail in Chapter 10. Two modulators are used and the subcarrier is applied to both of them, but the phase difference of the subcarrier signal between the two modulators is 90 degrees.

One modulator is fed with the R − Y signal and the other with the B − Y signal, as shown in Fig. 5.1. In actual fact, the R − Y and B − Y signals are weighted (e.g. altered in strength) before they are applied to the modulators and in the PAL system they are then known as V and U signals. The V and U signals and their weighting factors are considered in detail in Chapter 10. The complex signal resulting from this manner of submodulaton of the weighted R − Y and B — Y signals is called the *chrominance signal*, shortened to *chroma* in this book.

It is seen that the colour difference signals are fed to the modulators through filters passing signals from d.c. up to about 1 MHz. Remember the colour information only needs this small bandwidth. At this juncture it is sufficient to know that quadrature modulation applies to all the colour information to the main carrier in such a way as not to affect unduly the composite signal from the aspect of black-and-white sets.

The integrated signals form a complex signal containing components of both *amplitude* and *phase*. For chroma signal detection at the receiver, the missing

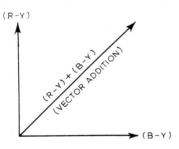

FIG. 5.2. *Showing how both the amplitude and the phase of the resultant chroma signal change when the amplitude of the R — Y and/or B — Y signals change.*

subcarrier has to be accurately reinserted so that it coincides with the phase and frequency of the subcarrier used at the transmitter. The luminance detector at the set, therefore, delivers the luminance signal plus the chroma signal and information on the phase of the suppressed subcarrier. On a monochrome transmission the detector delivers video signals to the luminance stages which then act in a similar manner to the video stages in a monochrome set, and the colour circuits are rendered inactive. It is noteworthy, however, that some sets are equipped with two vision detectors, one for the Y signal and the other for the chroma signal, which sometimes delivers the sound intercarrier signal also.

Figure 5.2 shows how two amplitude modulations, displaced by 90 degrees through quadrature modulation, give rise to both amplitude and phase changes when the amplitude of one or both modulations is varied. In this diagram the resultant represents the vector sum of the two chroma signal components. When the two are equal in amplitude, as shown in the diagram, the resultant is displaced equally (45 degrees) from each. Thus, change in amplitude of one or both of the signals causes both the angular position (phase) and the length (amplitude) of the resultant to change. In this way the colour information is identified by both phase and amplitude.

HOW THE PAL SYSTEM WORKS

Up to now PAL and NTSC are identical, but at this stage PAL uses a system to neutralise the effect of phase distortion on the chroma signal, either arising in the transmission or the equipment.

Changes in chroma signal phase represent a serious problem in the NTSC system because the signals are phase-sensitive, as we have just seen. This means that changes in the phase of the signals while they are travelling through space or through the transmitting or receiving system can change the reproduced colour. Colours can differ substantially from those transmitted, and even camera or transmission changes, having relative phase differences, can alter the hues. NTSC sets have a hue control allowing manual correction of this effect, but when phase changes are cyclic continuous re-adjustment is necessary, and viewers can have difficulty in establishing what is, in fact, the correct colour on the screen.

PAL solves the problem by alternating the phase of the R − Y signal at the transmitter line by line. The phase alternations are achieved by a switched inverter (PAL switch) in the circuit to the red modulator, and the inverter is activated by switching pulses at line frequency as shown in Fig. 5.1.

Figure 5.3 shows how the switching affects the colour vectors. Thus, the switching for one line of signal gives +(R − Y) and +(B − Y), for the next line of the same field −(R − Y) and +(B − Y), for the next line +(R − Y) and +(B − Y) and so on. The alternating R − Y phase is signified by the plus and minus signs.

Of course, both the R − Y and B − Y signals are transmitted on each line, but the phase of the B − Y signal remains the same on all lines, as governed by the nature of the chroma signal, and only the phase of the R − Y signal alternates over 180 degrees.

The PAL decoder is designed to respond to these R − Y phase alternations by rather clever means, as we shall see. In addition, PAL-D sets incorporate an electrical delay line. This 'stores' the colour information of one line of signal* and delivers it during the next line, along with the original signal of that line.

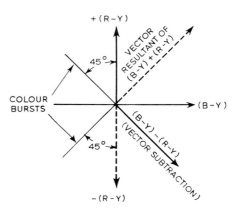

FIG. 5.3. *Showing how the R − Y signal phase alternates in step with the ±45 degree swings of the burst.*

The tube then displays lines of chroma signal corresponding to the electronic integration of the two lines of chroma signal.

PAL-D sets have the ability to cancel out electronically any phase error that may have crept into the system between the camera control and the receiver

---

* Although a line period is 64 μs, the PAL delay line introduces a delay of 63.934 μs because for the addition and subtraction of direct and delayed signals the delay must correspond to an exact number of chroma signal half cycles.

picture tube, thereby maintaining constant hue in spite of phase changes. The *amplitude* of the chroma signal is geared to the saturation of the colour, while the *phase* is geared to the hue. The reason for this is given in Chapter 10 which deals with encoding and decoding.

At this juncture, however, we can see that equal amplification or attenuation of the *two* chroma components *together* influences only the saturation; that is, a blue may appear a shade more or less blue. The hue remains correct in spite of this provided that the chroma signal or associated burst signal (see later) is not *phase* changed.

Changes in phase are much more dramatic because they change the colour on the screen. In the USA small phase changes sometimes occur when colour cameras are switched and when the viewer selects an alternative colour transmission. This calls for re-adjustment of the hue control, as just mentioned. Adjustment is often made on flesh-coloured tints in the picture or test card. It is worth noting here that some of the British PAL sets also feature a *tint control* for adjusting for the best flesh-colour on a card or picture (Chapter 2), but this is not the same as the NTSC hue control.

Changes in phase can also take place over poorly equalised video links, and in communal aerial and v.h.f. and h.f. relay systems. Multipath interference (i.e. 'ghosting') can also cause the effect. This is why a very good aerial system is essential for colour sets.

The phase-error cancelling attribute of PAL solves these annoying NTSC problems. Indeed, the ordinary hue control is not required in PAL-D sets. These sets can correct for errors in phase of up to 70 degrees, while PAL-S sets utilise subjective compensation, but can show lines ('Hanover blind' effect, see later) when the phase error goes beyond a few degrees.

Hue changes result from both positive and negative phase errors. Red in a picture, for example, would veer towards blue on a negative error and towards yellow-green on a positive one—depending on the magnitude of the error.

PAL-S SETS

PAL-S sets show colour errors in opposition on successive lines of a field. Let us suppose, for instance, that the camera is scanning a yellow picture element. In the absence of phase distortion, the shadowmask tube would display yellow on that element. However, with phase error in the system the display could veer from the correct yellow towards green when the error is negative and towards red when positive.

If the error veers towards the green, say, from the transmitted yellow, a PAL-S set will display the element veering towards green on one line of the field and towards red on the next line of the same field. Because the eye is insensitive to detail in colour, it will integrate the red and the green of that element on adjacent lines of the field and create the subjective sensation of the original yellow. This is really a wonderful idea in which the eye takes part in neutralising the actual *colour* error resulting from phase error.

Hanover blind effect

Under good conditions of reception, where phase errors are unlikely to be large, PAL-S sets can give good colour reproduction for less cost than PAL-D sets (because they are less complex), but where phase errors exceed a few degrees PAL-S sets exhibit an effect called 'Hanover blinds'. The effect was first observed in Hanover where the PAL system was developed.

This effect happens because the eye fails to be completely deceived when picture elements on successive lines of a field are substantially removed from the true colours as scanned by the camera; the eye then becomes aware that there are colour differences along successive lines.

Moreover, and this is an aggravating factor, since a complete frame of picture is composed of two interlaced fields, the nature of the PAL phase-alternations is such that the lines containing the errors of one field occur *adjacent* to the lines carrying the same error of the interlaced field. On one field, for instance, line 10 may carry elements in colour phase error, while line 12 (*not* line 11 because this is a line of the interlaced field of the frame) would then carry elements in reverse (corrective) colour phase. Now, line 11 of the interlaced field would be carrying elements in colour phase error, the same as line 10 of the partnering field, while line 13 would be carrying the elements in corrective colour phase, the same as line 12. The net effect is that the interlaced frame (complete picture) has adjacent pairs of lines with *each line* of the pair carrying the same colour information and *each pair* of lines carrying the elements in corrective colour. When the error exceeds a few degrees the eyes start becoming aware of horizontal lines across the picture, and the lines become disturbing as the phase error increases.

PAL-D SETS

PAL-D sets are not troubled by the 'Hanover blind' effect when adjusted correctly because the correction is electronic, not subjective. The signals corresponding to the lines in corrective phase and in phase error are 'mixed' before they are translated into displays on the tube. This is where the PAL electronic delay line comes into effect. Each line of a field is delayed by the time taken by one line and then released for mixing with the next line of the same field. Each line of signal received by the tube thus contains the information appropriate to the two lines of alternated colour phase. One line of signal is fed straight into the 'tube circuit', while the signal of the phase-alternated line, fed in at the same time, is derived from the PAL delay line. Details are given in Chapter 10. Figure 5.1 (the block diagram of the PAL transmitter) also shows a burst-pulse generator. This is an oscillator which is phase-locked to the subcarrier generator, and the signals that it delivers eventually control the reference generator in the set.

Since the subcarrier is suppressed at the transmitter, chroma demodulation requires the subcarrier to be reinserted. Moreover, it must be accurately 'locked' to the phase of the suppressed subcarrier to keep the reproduced colours true.

There must also be provision for alternating the red colour-difference signal at the receiver in synchronism with the PAL alternations at the transmitter.

SWINGING BURSTS

Two types of 'colour synchronisation' are required by the PAL system. The first is to phase-lock the receiver's reference signal, and the second is to ensure that the red detector at the receiver switches in step with the $R - Y$ alternations at the transmitter.

These two processes are handled by the signals delivered by the transmitter's burst generator. These are called *colour bursts*, and they are transmitted during the back porch periods of the line sync pulses (see Fig. 1.2). The PAL switching of the red detector is synchronised by the colour bursts being alternated by ± 45 degrees relative to the $B - Y$ component in step with the $R - Y$ signal alternations at the transmitter, and for this reason the colour bursts of the PAL system are known as *swinging bursts*. The action is shown in Fig. 5.3. More information about this is contained in Chapter 10.

The remainder of Fig. 5.1 is concerned with the PAL switching of the $R - Y$ signal in step with the bursts, and the addition of the sync, 'black level' and bursts to the composite signal at the correct places on the waveform. The PAL transmitter is in reality more complex than the diagram implies.

Incidentally, the delay line at the transmitter in the Y channel simply increases the time taken by the Y signal so that it arrives at the adder at the same time as the colour signals. Because the bandwidth of the colour channels is less than that of the luminance channel, colour signals take a small fraction of a second longer to pass through them, and the luminance delay compensates for this.

ADDING COLOUR TO THE COMPOSITE SIGNAL

We must now see how the chroma signal is added to the luminance signal within the standard 625-line vision channel. The line-by-line creation of a monochrome picture concentrates much of the signal energy into discrete, narrow bands at multiples of the line frequency. This produces relative low energy gaps between the high energy monochrome bands, and these accommodate the lower definition colour information. The idea is illustrated in Fig. 5.4.

SUBCARRIER FREQUENCY

The subcarrier must, of course, fit into the video-frequency spectrum, and it is carefully worked out to give the lowest visibility dot-pattern on black-and-white sets. If the dots produced by the subcarrier are in orderly rows they trouble a monochrome picture more than when displaced. To ensure the least interference the subcarrier frequency (f_{sc}) is related to the line scanning frequency according to the expression

$$\text{line frequency} = \frac{f_{sc} - \frac{1}{2}f_{field}}{284 - \frac{1}{4}}$$

For a 625-line picture with a field frequency (f_{field}) of 50 Hz, the subcarrier frequency works out to 4·43361875 MHz $\pm$ 1 Hz. This is the frequency adopted

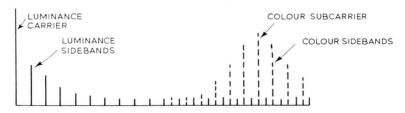

FIG. 5.4. *Chroma information is carried in the low-energy gaps between the luminance information.*

for the British PAL System, and at the receiver a crystal oscillator is used to generate it.

PAL RECEIVER

Now let us look at the stages in a PAL receiver. The video stages are shown in Fig. 5.5. Figure 5.6 shows that the chroma signal can be obtained from the main detector along with the Y signal (a) or from a separate detector (b). The diagrams indicate that from the tuner to the detectors a colour set is similar to a mono-chrome one. Dual standard and 625-line-only models are in current use, but the latter type will eventually replace the former when the v.h.f. channels are phased out; it is the single-standard type that is indicated in the diagrams.

Figure 5.5 shows that the composite signal is fed to the luminance amplifier via a delay line and a notch filter. This signal—the Y signal—is applied to the tube cathodes in many sets and it is responsible for providing the 'brightness information' and detail. The notch filter gives a sharp attenuation at the chroma subcarrier frequency which deletes all the chroma information from the Y channel, thereby eliminating dot-pattern interference. The delay line* 'slows down' the wideband luminance signal so that it arrives at the picture tube at the same time as the colour signals via the chroma and colour-difference circuits. A delay line fault has the effect of displacing the colour from the luminance on the screen. Poor termination of the line can cause 'ghosting', and the same symptom can arise from misalignment in the chroma and/or luminance amplifiers, as this can alter the time taken by the signals in passing through them.

The chroma signals with the bursts are fed through the chroma bandpass filter and then to the PAL decoder. This results in the production of the original R − Y and B − Y signals.

* The value depends on circuit bandwidths.

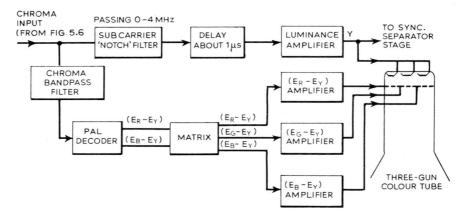

FIG. 5.5. *Block diagram of the video stages. Note that Y delay depends on luminance frequency/- bandwidth characteristics. Delay generally less than 1 μs, typically 600ns. (See Fig. 5.6 below and Fig. 8.11. for example.)*

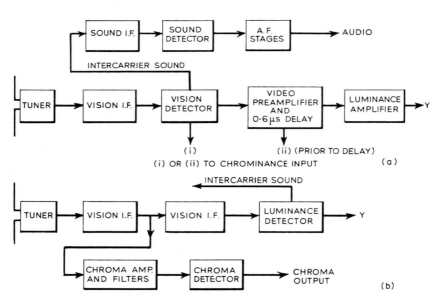

FIG. 5.6. *Chroma signal can be obtained either from the main detector (a) or from a separate chroma detector (b). Sometimes the intercarrier signal is taken from the chroma detector.*

DERIVING G — Y

The G — Y signal which was not transmitted is obtained from a matrix follow-ing the decoder, at the input of the colour-difference channels. It will be realised that the R — Y and B — Y signals must embody the missing G — Y signal since they carry information on Y. The G — Y signal can be obtained from a matrix because of the relationship $Y = 0.3R + 0.59G + 0.11B$. It can be shown (page 129, Chapter 8) that $G - Y = -0.51(R - Y) - 0.19(B - Y)$.

The matrix network is designed to invert both the R — Y and B — Y signals (that is, to change their signs), to adjust their respective levels to the ratios given and finally to add them, resulting in an output of G — Y.

Note that the chroma detectors and matrix deliver the three colour-difference signals only. The original R, G and B *primary colour* signals are obtained by subtracting −Y from each of the colour-difference signals. This is often accomplished by the matrix action of the picture tube guns. For instance, because the −Y signal is applied to the three cathodes together and the three

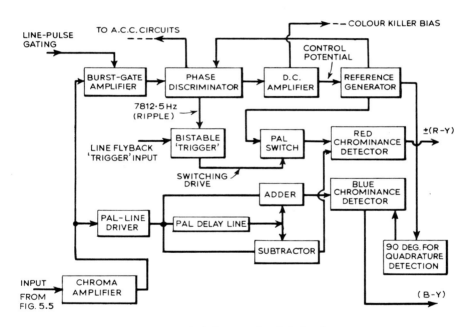

Fig. 5.7. *Block diagram of PAL-D decoder.*

colour-difference signals to the appropriate grids, each gun is modulated by the corresponding primary colour signal. Some sets use a circuit matrix, instead of the tube, for deriving each primary colour signal from its colour-difference signal.

A very important section of the PAL receiver is the decoder, and the stages of this are revealed in the block diagram of Fig. 5.7. Signals from the chroma bandpass filter (Fig. 5.5) are fed to the chroma amplifier and simultaneously to the burst-gate and amplifier. This 'gate' opens only when colour-burst signals are present on the transmission, passing them to the phase detector. Details of this and the other parts of the receiver are considered in Chapters 8, 9 and 10.

Phase-locking the subcarrier

Recovery of the colour-difference signals is only possible by reinserting the suppressed subcarrier in a phase that matches exactly that of the subcarrier at the transmitter. A crystal-controlled subcarrier generator, or *reference generator* as it is called, is used at the receiver. This phase locks to the colour bursts, and they can then, in fact, be considered as a continuous extension of the subcarrier at the transmitter. Phase locking is achieved by a phase detector. This receives the colour bursts and reference signal. When these two signals are phase coincident, there is no corrective output. However, should the phase of the reference signal tend to deviate from that of the bursts, the phase detector yields a control potential. This potential is fed to a d.c. amplifier, and thence back to the reference generator to bring it into phase with the bursts.

This is achieved by a capacitor-diode across the crystal of the generator. The value of the diode changes as the reverse bias across it is altered, and the control potential from the phase detector represents this bias. The action is similar to that of flywheel-controlled line synchronising.

The controlled reference signal is fed to the red and blue detectors. A 90 degree phase-shift network is interposed in the feed to one of the detectors to secure the condition needed for detection of quadrature modulation. In the feed to the red detector, however, is interposed the PAL switch. This alternates the red colour-difference signal line-by-line in step with the alternations at the transmitter.

The phase modulation on the bursts due to the ± 45 degree alternations appears as a half line-frequency signal (7·8125 kHz, but generally given as 7·8 kHz) at the output of the phase detector. This signal, commonly called 'ripple signal', is filtered and amplified and then used to *identify* the switching phase for the red detector. Since the 'swings' of the bursts are equal either side of the B — Y chroma component, and since a long time-constant is included in the control circuit, the reference generator stabilises at the correct average phase.

The red detector PAL switch is operated by a bistable triggering circuit. This uses a pair of triode valves or transistors arranged rather like a multivibrator but *without* the normal oscillation sustaining feedback. Both halves of the circuit are thus stable (hence *bi*stable) with one valve or transistor conducting while the other is switched off. To cause the 'switched-off' valve or transistor to switch on and the 'switched-on' one to switch off, a pulse has to be applied to one side of the circuit. Each pulse causes a changeover, and this action operates the PAL switch.

The bistable circuit is triggered by line-frequency pulses, so that the PAL switch operates line-by-line. It is possible, though, for the PAL switch to give

+(R − Y) while the transmitter is sending −(R − Y). This would severely degrade the colour reproduction as can well be imagined. The correct line at the receiver is identified by the half line-frequency ripple signal processed, as just described.

This signal is applied to the other side of the bistable circuit and, provided that the switching corresponds to the correct line, the ripple simply helps the line pulse to trigger the bistable circuit. However, if the lines are unmatched, the pulses oppose and the circuit fails to trigger on one count, thereby automatically restoring switching synchronism. Because the ripple signal identifies the correct phases of the red lines it is called an identification signal in this context—or simply 'ident pulses'.

The PAL adder and subtractor are fed with chroma signal by two paths: one direct from the delay line driver amplifier and the other via the delay line. The PAL delay is close to 64 μs, which is equal to the time taken by one complete (625 standard) line. This means that the adder and subtractor receive two lines of signal, but one is one line late.

Signals of two consecutive lines are thus both added and subtracted. Conditions are arranged so that a change in the red signal cannot alter the adder output nor can a change in the blue signal alter the subtractor output; the result of this is that phase error developing in either one of the chroma signals has no effect on the other.

Owing to the nature of this form of detection, a phase change in one of the chroma signals resolves as an *amplitude* change after detection, and because amplitude affects the saturation and not the hue, a phase change in PAL-D sets can only produce a change in the saturation of the colours displayed at any instant. The principles involved are fully investigated in Chapter 10, and other chapters explore all the refinements of the PAL colour set, which we have not been able to consider in this bird's eye view of the system.

CHAPTER 6

Purity and Convergence

LET US FIRST clarify our thoughts on purity and convergence. If we think of one non-deflected beam, we can understand how this can be lined up—through shifting its effective starting point by a purity magnet on a 'purity axis'—so that it shoots through several adjacent holes in the shadowmask and hits only phosphor dots of its 'own' colour.

We can also understand that when deflection is applied to form a raster, provided that the right deflection point is found on the purity axis, the beam will still hit only the right dots. This lining up of deflection centre, holes and dots is built into the tube—within remarkably close manufacturing tolerances.

So much for purity of one beam. The other two beams can be set up similarly, giving rasters in two more pure colours. With three beams of the right relative strengths, a white raster will be seen.

Suppose signals are now supplied. The chances are that three images all slightly displaced from each other—and in mixed-up hues—will be seen. Why? Because the beams are not converged into a single scan spot.

We can visualise the three static beams hitting the mask somewhere near the centre but at three different points. Clearly, by applying static magnetic fields we can bring them into convergence—and without affecting the purity adversely, provided that the necessary corrective deflection occurs at the deflection centre on the purity axis.

We then have pure *and* converged beams at the centre. When raster deflection is applied (by a single set of line and field coils), the fact that the three beams are displaced from each other at the deflection centre (the guns being grouped around the tube axis) means that they are affected slightly differently. As the beams leave the zone of static convergence some different form of convergence control is needed.

This is obtained by modifying the overall line and field deflection forces for each beam individually by 'dynamic convergence' coils—a pair for each beam—carrying suitably shaped correction currents.

DEGAUSSING

We must also remember that stray magnetic fields due to parts of the receiver becoming magnetised, and even the Earth's field itself, can affect control of the three beams. Before starting purity adjustments, therefore, a set must be demagnetised (or degaussed).

As explained earlier, most receivers have automatic degaussing. This means that an inbuilt degaussing coil receives a current, which decays slowly each time the set is switched on. However, when the set is first installed and if it is moved in the room it is likely to require degaussing with an external coil, even though it has built-in degaussing.

Other things must also be checked before adjusting for purity. These are the picture size and tilt (tilt should be checked prior to locking the scanning assembly after establishing the correct deflection centre), centring, linearity and focus. One should also make sure that the convergence is reasonably well in adjustment before tackling purity. There are electrical as well as physical adjustments for convergence—and this is a matter we shall come to later.

Automatic degaussing

Automatic degaussing is accomplished by a pair of degaussing coils located under the magnetic shield of the picture tube cone (Fig. 6.1) and a control circuit which energises the coils each time the set is switched on, followed by a gradual

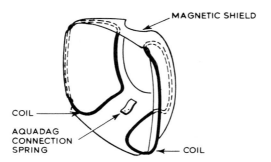

FIG. 6.1. *Automatic degaussing coils underneath the tube's magnetic shield. The number of turns of wire making up the coils and the diameter of the wire depend on the nature of the control circuit and power supply.*

decay to zero. The automatic degaussing circuit also operates in some sets when a change in line standard is made (e.g. when changing from a 405-line to a 625-line channel and vice versa).

A variety of control circuits have been evolved for the control and a very simple example is given in Fig. 6.2. Here a thermistor with a positive temperature co-efficient is connected in series with the coils and the 'auto switch', which is in the 'on' position when the set is switched on. The p.t.c. thermistor has a low resistance when cold (about 80 Ω) and when mains current flows through it and the coils it warms up, increases in resistance and gives a reducing current in the coils and hence a reducing magnetic field. After a few seconds, when the current is at a minimum, a bi-metal relay action or built-in delay opens the 'auto switch' which disconnects the coils from the mains supply.

63

A more sophisticated arrangement that avoids the use of an 'auto switch' is shown in Fig. 6.3. At switch-on the p.t.c. thermistor is cold, its resistance is low and the voltage across the lower network is, therefore, high. A high current thus flows in the coils and the series voltage-dependent resistor (v.d.r.) and this produces a maximum magnetic field. As the p.t.c. thermistor warms up, so its

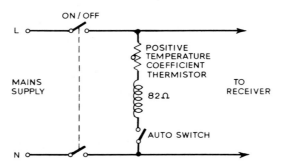

FIG. 6.2. *Simple automatic degaussing control circuit, requiring an automatic switch which is 'on' when the main on/off switch is closed, and which automatically opens a few seconds later, when the thermistor has increased in value.*

resistance increases and the voltage across the lower network falls. Current in the coils falls because of this, but it is further reduced by the v.d.r. also increasing in resistance (because of the reducing voltage across it). The thermistor current then bypasses the degaussing coils by flowing through the 680 Ω shunt resistor. In this manner, therefore, the switch-on degaussing current is high, falling gradually to zero as the thermistor and v.d.r. increase in resistance, terminating

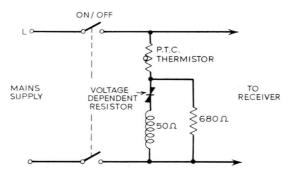

FIG. 6.3. *Automatic degaussing control circuit with p.t.c. thermistor and v.d.r., not requiring an automatic switch.*

in a stabilising current flowing through the thermistor and the 680 Ω resistor.

There are various other automatic degaussing control circuits whose principles are based on the two circuits described. The energising current for some automatic degaussing circuits is derived from a step-down winding on the mains transformer.

The automatic degaussing module of the British Radio Corporation (BRC) 2000 series sets is shown in Fig. 6.4. Note the large, disc-type v.d.r. and the two pairs of thermistors.

External degaussing coils

An external degaussing coil is required to remove residual magnetism from metal items inside and sometimes outside the set, which are outside the influence of the inbuilt coil.

Commercial degaussing coils are readily available, but a coil can be made if required by winding 800 turns of 24 s.w.g. enamelled copper wire on a 12 in.

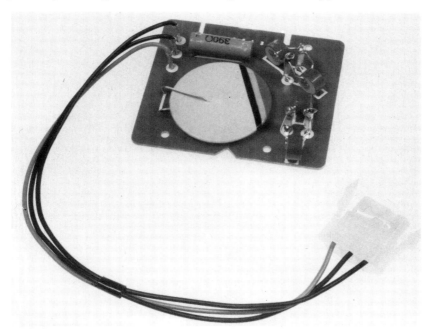

FIG. 6.4. *Automatic degaussing module of the BRC series 2000 receivers. Notice the v.d.r. and thermistors, the latter arranged in pairs (courtesy, British Radio Corporation Ltd.).*

diameter mandrel. This should be double-wrapped with Empire or PVC tape to provide adequate insulation. About five yards of mains cable is sufficient to connect the coil to the mains supply. This sort of coil will yield a field of 75 G measured $\frac{1}{2}$ in. from the exterior of the winding.

Manual degaussing procedure

1. Connect coil to the a.c. mains supply.
2. Move the coil slowly about the faceplate of the picture tube and over the

top, bottom and sides of the cabinet—*but never at the back*—for about one minute to remove any isolated magnetism. (Note: the set need not be switched off.)

3. Finish the exercise with the coil over the centre of the screen and withdraw it slowly to a distance of at least 8 ft before disconnecting the mains supply. (Note: This also means 8 ft from any other colour set, whether operating or not.)

The coil can also be used to demagnetise any external ferrous objects, such as a metal frame piano, radiators and so forth located near the receiver. However, good installation practice always ensures that the receiver is sited well clear of such objects.

PRINCIPLES OF PURITY

The scan coils work the same way as in monochrome sets but have three beams to deflect simultaneously instead of one. In monochrome sets the practice is to push this assembly hard against the tube flare to avoid corner shadowing, but in colour sets the coils have to be positioned to satisfy colour purity requirements. The assembly is provided with a slot adjustment for positioning and a wing-nut for locking.

Pure red, green and blue rasters are produced only when the electron beams strike all their associated colour phosphor dots on target (Fig. 6.5a). An error in the angle of approach of a beam to the shadowmask will result in a partially 'off-target' strike and often the illumination of phosphors of the 'wrong' colour (Fig. 6.5b). Since the approach angle of the beams is influenced by the position along the neck where the scanning deflection occurs (Fig. 6.5c), it follows that the neck position of the scan coils affects the purity.

Facilities are provided to bias-off the electron beams independently so that purity can be judged on red, green and blue rasters separately. The red first, and then the other two colours, are checked. The degree of purity is indicated by the extent a one-colour raster is contaminated by other colours. Ideally, each raster should be illuminated solely in its colour, and in even brightness, over the whole area of screen. This ideal is rarely achieved in practice. Tubes are set up for satisfactory results when seen from a normal viewing distance.

SWITCHING OFF THE BEAMS

British colour receivers have switches marked red, green and blue for cutting off the beams of one gun or more as required for purity and convergence adjustments and for testing and fault-finding.

These switches are shown on the dynamic convergence chart in Fig. 6.25 (explained later) and they can also be seen on the convergence module in Fig. 6.27.

In most sets the switches interrupt the supply to the first anodes of the tube guns. This is shown in Fig. 6.26, where S2, S3 and S4 operate respectively in the blue, green and red guns, putting the first anodes at 'earth' in the 'off' positions.

Alternatively, any required gun can be muted by connecting a 100 kΩ resistor between its control (modulating grid) and a convenient 'earth' point.

The three guns are displaced equally by 120 degrees within the tube neck and inclined slightly towards the axis so that the beams converge to a point at the

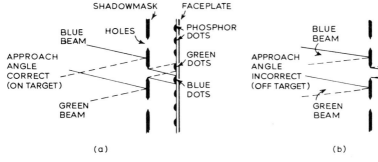

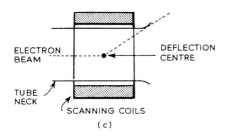

FIG. 6.5. *Pure red, green and blue displays are obtained only when the three beams arrive at the shadowmask at the correct angle to give 'on-target' hits at the phosphor dots (a). Correct colour dot may be only partly lit and adjacent incorrect colour dot lit, too, with incorrect beam angle (b). To ensure the correct angle, the beams must be deflected at the right position within the scan assembly (c).*

shadowmask. Diagrams (*a*) and (*b*) in Fig. 6.5 show only two of the three beams for simplicity of drawing. While manufacturing tolerances are geared to minimising the error as in Fig. 6.5(*b*), it is impossible by mechanical means to guarantee on-target phosphor-dot hits even when the beams are travelling down the centre of the tube.

An important point to note is that the purity at the centre of the screen is not affected by the position of the scanning coils. This is fairly obvious when one comes to think about it, since there is no deflection under this condition.

SCREEN-CENTRE PURITY ADJUSTMENT

Purity is, therefore, first established in the middle of the screen by adjusting the purity magnet on the tube neck. Then, away from the centre towards the edges, it is adjusted by positioning the scan coils along the neck.

To simplify purity magnet adjustments, the scan coils are released and slid along the locating slot towards the tube base or towards the tube flare. This destroys the purity towards the edges and may put shadows on the corners, but this does not matter at this stage because the aim now is to secure the best possible purity patch in the middle of the screen—red first and then the other colours —taking no particular heed of what is happening round the edges.

The purity adjustment consists of a pair of ring magnets designed to fit the tube neck and which rotate relative to each other. Each ring carries tabs or mouldings to facilitate rotation and reveal how far it is turned relative to the other. Adjustment is similar to monochrome set picture centring, as explained in Chapter 4.

The amount of purity adjustment required depends on any external fields but mainly on the tolerance spread of the tube itself. Sometimes little or no correction is needed and this is why it is necessary to be able to cut the field across the neck almost to zero. The purity field must be uniform across the neck so that the three beams are equally affected.

It should be noted that there is some small interaction between the purity field and the static and dynamic convergence fields. The purity field tends to bend the three beams together prior to them entering the scanning coils. Since the purity field can be adjusted both in intensity and polarity the exact angle of approach for on-target hits at the middle of the screen can be established.

The three beams together are moved in a circular cross-section path when the two rings are turned in the same direction together and radially across the neck when there is relative movement of the two magnets. That is, spreading or closing of the tabs.

From this description and by studying (a) and (b) in Fig. 6.5 it will be appreciated that sufficient flexibility is given by the simple purity magnet system to correct purity errors in the middle of the screen.

SCREEN-EDGE PURITY ADJUSTMENT

There is now the rest of the screen to purify. The ultimate aim is to obtain purity over the whole raster of each colour in turn. This is where adjustment of the scan coils comes in.

It will be remembered that the assembly was moved away from the tube deflection centre prior to adjusting the purity magnet. It is necessary now to slide it back towards its original position. As this is done, purity spreads from the small patch in the middle of the screen towards the edges of the raster. The position along the neck corresponding to the best overall raster purity is selected and the wing-nut tightened.

This step is generally carried out on the red raster (the green and blue guns switched-off) and checks are made afterwards separately on the other two colours. It might be found that there is sufficient tolerance on the red adjustment to allow slight readjustment for purity on the three colours. A compromise adjustment is then established for the best possible purity on all three colours. If it is impossible to secure satisfactory purity, a check should be made on the purity magnet adjustment, after sliding the scan coils away from the deflector centre as before.

While the purity magnets correct the approach angle of the three beams towards the middle of the shadowmask as we have seen, correction is held at all deflection angles only when the deflection centre has been accurately established by sliding the coils on the neck. When the deflection centre is wrong, the effect is that the beams progressively strike their appropriate colour phosphors off-target (this shows up as loss of brightness) until, towards maximum deflection, they are striking phosphors of incorrect colour.

Thus, when adjustment is being made for the colour patch in the middle of the screen, the remainder of the screen will display different hues. The exercise is to *centre* the patch of *correct* colour within the background sea of different hues.

A pure tube will display a white raster (when all three guns are delivering beams of correctly proportioned current), and this is the final test for purity. Pattern generators have a 'white raster' position for purity checks.

Areas of colour tinting on the white raster indicate that the illumination from the individual colour rasters is uneven over the whole screen; also, of course, the light output is below the general level where the shading occurs. When working on single-colour rasters it is not easy to discern variations of illumination due to the beam being slightly off-target.

PRINCIPLES OF CONVERGENCE

While purity involves corrective deflection of the three beams together, convergence adjustments are made on each beam individually. Small ceramic permanent magnets control the static convergence and coils control the dynamic convergence.

The permanent magnets are adjustable in such a way that the field transferred to the tube pole-pieces (Chapter 4) can be increased or decreased. Sometimes this is achieved by the magnets being movable from the assembly poles or by the magnets rotating within ferrite sockets.

Sometimes a direct current in the dynamic coils partly or wholly produces the static convergence fields. This is so in some dual-standard and single-standard sets, the direct current being produced sometimes by diodes rectifying the line timebase convergence signals.

The permanent magnets converge the three beams correctly towards the middle of the screen, while the electromagnets apply a changing field to maintain convergence as they sweep towards the edges of the screen. Hence the terms *static* and *dynamic* convergence.

Correct convergence occurs when the three beams produce a close cluster of red, green and blue light at any scanning point on the screen, representing the scanning spot when viewed at a distance. If one beam is incorrectly converged, it will produce a cluster of its own colour light misplaced from the main cluster when the tube is 'pure'.

In spite of misconvergence, purity is not necessarily lost. That is, the beam which is misconverged may still produce a cluster (about three) dots of its correct colour. The idea is shown in Fig. 6.6. Here the blue beam is too far down but still giving blue light. On a monochrome signal this would appear yellow (red plus green equals yellow) and displaced blue.

The convergence assembly and tube neck are shown in cross-section at (*a*) in Fig. 6.7 and in reality at (*b*). The magnetic fields produced by each arm of the assembly cause an up and down movement (towards and from the tube axis)

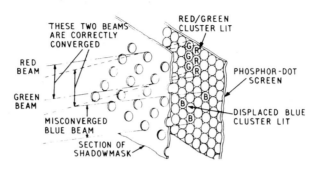

FIG. 6.6. *It is possible to secure correct purity with misconvergence, as shown here. The blue beam is lighting the blue phosphors, but these are displaced from the mains group lit by the red and green beams.*

of the associated beam. Magnetic shielding prevents the fields from affecting the two beams they are not associated with. Thus, by adjusting the permanent magnets at the top of each core or arm the associated beam is moved as shown by the arrowheads. The misconvergence just mentioned, therefore, could be corrected by adjusting the blue convergence magnet—the one at the top of the tube neck—to move the beam a little away from the tube axis.

The green and red beams move along lines that are inclined 30 degrees from the horizontal, and viewing along the neck from the base end, green is on the left and red on the right.

Misconvergence

Misconvergence does not show up on a white raster without picture modulation. When the tube and the beam intensities are correctly proportioned, a white raster will always resolve, irrespective of convergence error. Convergence relates to the correct registration of the three colour images. Thus, we must have modulation on the three rasters to observe misconvergence.

A monochrome picture shows up misconvergence by the detail of the picture being surrounded by colour. This is called *colour fringing*. Indeed, running a colour set on a monochrome transmission shows up colour adjustment shortcomings dramatically.

The transmitting authorities radiate colour Test Card F (Chapter 11), and while this highlights convergence maladjustments, one has to be highly skilled to correct the adjustments on the test card alone without recourse to a crosshatch and dot generator. Switched to dot pattern, such a generator (Chapter 11) produces a uniform pattern of dots on the screen. It will also be switchable to a crosshatch (series of vertical and horizontal lines). And some models produce a colour encoded signal, too. The picture information is modulated on to a u.h.f. or v.h.f. carrier, making it a simple matter to connect the instrument to the set on either standard. Pattern generators and other instruments are considered in Chapter 11.

A set reasonably well adjusted for picture width, height, shift, focus and linearity, in which the tube is pure, will display its convergence performance in

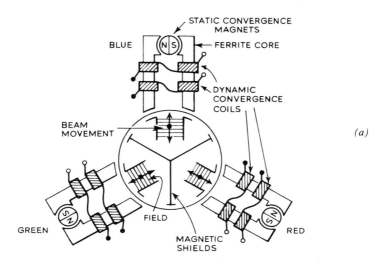

(a)

(b)

FIG. 6.7. *Each beam has a static (and dynamic) convergence field to adjust it. The dynamic convergence coils comprise one pair for field and one pair for line. (a) Cross-section of system and (b) assembly on tube neck, with the other neck components (courtesy, Mazda Colour TV Tubes Ltd.)*

the middle of the screen when a dot-modulated generator signal is applied to it and its edge-of-screen convergence performance best when the modulation is changed to crosshatch, but it is possible to use a crosshatch pattern throughout.

71

Middle-of-screen convergence is regulated by the static field adjustments (permanent magnets) and the edge-of-screen convergence automatically by current waveforms in the special windings. Static misconvergence is revealed by colour fringing round the dots in the middle of the screen and really severe misconvergence by obvious displacement of the red, green and blue elements of the dots (Fig. 6.8b).

Dynamic misconvergence, on the other hand, is indicated by the lines of the

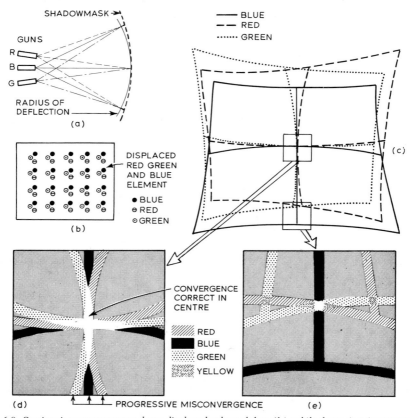

FIG. 6.8. *Static misconvergence produces displayed coloured dots (b), while dynamic misconvergence (d) gives progressive displacement of lines from the centre of the screen (on crosshatch display), dividing the three colours. The displacement of the three guns from the tube axis (a) produces different pin-cushion distortions on red, green and blue (c). The green distortion is a mirror image of the red, while the blue is similar but turned through 90 degrees. The effect in the centre of the screen is shown at (d) and at the edges of the screen at (e). The effects are exaggerated here for clarity.*

crosshatch progressively dividing into separate red, green and blue lines towards the edges of the screen (Fig. 6.8c).

HOW TO ADJUST STATIC CONVERGENCE

Static convergence is tackled first and then dynamic convergence by adjusting the strength and shape of the current waveforms in the dynamic coils. When there is bad misconvergence, it is best to switch off the blue gun so that only the red and green elements of the dots are displayed. The red and green static magnets can then be adjusted until the two dots merge and give yellow light. The magnet adjustments should be balanced as far as possible about their centre of adjustment. The dots, as given by the signal modulation, should be those in the centre of the screen.

The blue beam can be restored, and the exercise is then to get the blue elements to merge with the yellow dots in the middle of the screen, to give white dots. It will probably be impossible to secure this condition with the blue radial magnet

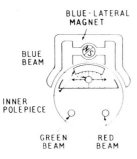

FIG. 6.9. *The blue lateral assembly affects only the lateral shift of the blue beam.*

alone, for the green and red elements might well have been converged to give yellow light to the right or left of the vertical line along which the blue element of the dot will move when the radial magnet is adjusted.

It is necessary to be able to shift the blue beam laterally as well as radially, and this is where the blue-lateral assembly comes in. Figure 6.9 shows a cross-section of this assembly on the tube neck, and the arrowheads indicate the way in which the blue beam moves when the field is changed in intensity and polarity.

FIG. 6.10. *While the radial convergence fields allow convergence of the green and red beams, the blue beam requires a lateral shift, too, to allow it to converge with the green and red beams, as shown here.*

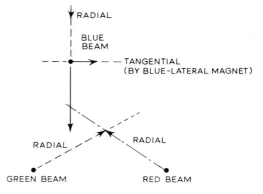

Figure 6.10 sums up all this and shows how the three beams are shifted radially by the main convergence assembly and the blue one laterally by the blue lateral assembly. Clearly, then, if convergence of the green and red elements puts the resulting yellow spot off the line of the blue radial movement, the blue beam can be shifted laterally and finally radially until it adds to the yellow spot to produce a white spot. This completes the static convergence.

However, in practice there is often some interaction of the originally merged green and red elements when the blue-lateral magnet is adjusted, making it necessary to repeat all the adjustments until the middle-of-screen dots are completely free from colour fringing. The dynamic convergence coils create changing magnetic fields to increase the distance along the three beams to the point where they converge, as their deflection angle increases. In other words, the coils compensate for the relatively small curvature of the screen and shadowmask compared with the larger radius of deflection. They also compensate for the displacement of each electron gun relative to the tube axis.

The effect on the display of a monochrome tube is the well known pin-cushion distortion, as shown in Fig. 6.11, but since there is only one beam to worry about correction is not unduly difficult and convergence problems do not exist. By the careful positioning of small magnets round the scanning assembly or tube flare monochrome pin-cushion distortion can be almost

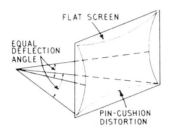

FIG. 6.11. *Showing how a single raster suffers pin-cushion distortion due to the radius of the screen failing to coincide with the radius of scan of the electron beam.*

completely eliminated, but such magnetic correction cannot be used relative to the shadowmask tube since the fields would disturb the convergence and purity.

THREE PIN-CUSHION DISTORTION CHARACTERISTICS

Each of the three uncorrected rasters of a colour tube has its own pin-cushion distortion characteristic because each gun is displaced differently from the tube axis. Figure 6.8(*c*) shows these distortions on the red, green and blue rasters. It will be seen that the gun displacements introduce degrees of trapezoidal distortion to the basic pin-cushion distortion, and that the green distortion is a mirror image of the red, while the blue is similar but turned through 90 degrees.

The need for correction is thus clearly revealed and correction is provided by the dynamic convergence which, from the display point of view, can be regarded as a scheme for pulling the three rasters into a common shape, so that over the whole screen area they fall almost exactly on top of each other. Any remaining pin-cushion distortion is then eliminated on many sets by interconnecting the field and line scanning currents by means of a device called a 'transductor' (see Chapter 7, page 95). A 'dodge' adopted by BRC in the single-standard 3000 series clears residual pin-cushion distortion subjectively. It consists of mounting the picture tube with the blue gun *down* instead of in the more usual 'up' position. The arrangement puts the greatest pin-cushion error along the bottom of the tube face where it is less noticeable from the viewing position. Normal dynamic (and static) convergence is employed of course.

The field and line convergence circuits are energised from the field and line timebases, and the circuits then supply specially tailored current to the three pairs of windings on the convergence assembly. The basic principle is shown by the block diagram in Fig. 6.12, in which it will be noted that the blue-lateral coil is energised only from the line convergence circuit, while the radial assembly is

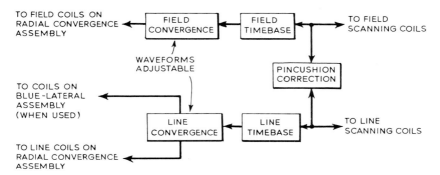

FIG. 6.12. *Basic principles of dynamic convergence and pin-cushion correction. Convergence correction currents are obtained from the line and field timebases, and residual pin-cushion distortion is sometimes corrected by 'crossmixing' the timebase signals.*

energised from both line and field circuits. Each arm has two sets of coils and each set is in a series-aiding configuration (see Fig. 6.7).

Figure 6.8 indicates that the distortion towards the sides of the rasters is of the nature of a parabola, and because of this it is possible to correct the error by applying a parabolic current to each of the dynamic convergence coils. The trapezoidal distortion, resulting from the displaced and slightly inclined guns (Fig. 6.8a), however, requires a different correction current, which is of a sawtooth nature. Thus, the convergence coils receive a 'mix' of parabolic and sawtooth currents.

Optimum correction over the entire screen calls for careful regulation of the amplitude and the mixing of the parabolic and sawtooth currents, and each convergence coil has to be individually adjusted in these terms, leading to twelve or more presets on a single-standard model and up to double that number on a dual-standard model.

DYNAMIC CONVERGENCE SYSTEMS

Figure 6.13 shows the problem—the manner in which the distance of the point of beam convergence from the centre of deflection increases as the beams sweep to the edges of the screen. Clearly, the convergence correction fields, produced by currents in the dynamic convergence coils, must shift the convergence point 'outward' as the beams are deflected.

Basically, there are two methods of obtaining the currents:

1. Applying voltage across the convergence coils, in which case these and the scan coils are effectively in parallel.
2. Passing the actual scan current through the convergence coils, in which case these are in series with the scan coils.

With (1) the convergence coils must have a reasonably high inductance to avoid upsetting the loading on the timebase while with (2) their inductance must be low, for the same reason.

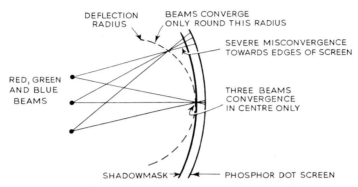

Fig. 6.13. *The problem of dynamic convergence is here shown in detail. Since the radius of deflection of the beams fails to coincide with the radius of the shadowmask and phosphor-dot screen, convergence occurs only at screen centre (see also Fig. 6.8). Misconvergence occurs progressively towards the edges of the screen, and the nature of the resulting pin-cushion distortion is modified due to the three guns being displaced from the tube axis.*

Some receivers use circuits derived from those developed by Mullard and Thorn-AEI. The Mullard circuit adopts scheme (1) and the Thorn-AEI scheme (2). There are also variations of these two themes.

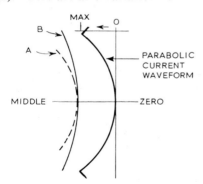

Fig. 6.14. *This diagram shows how a parabolic current in the dynamic convergence coils will correct the misconvergence from the centre of the screen outwards, by causing the beams to remain in convergence along the radius of the screen B in spite of the greater radius of scanning A. However, slight deviation from a true parabolic waveform is necessary to compensate for the displacement of the guns from the tube axis. Suitable deviation is provided by the addition of sawtooth waveform.*

Figure 6.14 shows the shift of convergence point, which is parabolic, together with a waveform of the kind needed for the correction current. Figure 6.15 is an oscilloscope picture of three actual current waves from a field convergence feed circuit.

'Plain' parabolic waveforms are not good enough, however, for the reason that the three beams come from guns offset from the axis of the tube. Take the blue gun, for instance, located between and above the red and green guns: the beam of this gun is longer when reaching the bottom of the screen than the top (see Fig. 6.16). With the red beam, however, the length is greater when deflected to the right than deflected to the left; conversely with the green.

Correction of the distortion produced by gun displacement is possible by displacing in time the zero point of the corresponding parabolic convergence current and giving the parabola a bit of tilt. These factors are brought out in Fig. 6.17 (page 78) for the field (a) and for the line (b) convergence currents. Diagram (c) shows the required deviations for convergence.

Fortunately both displacement and tilt can be applied to a basic parabolic waveform by adding a little sawtooth waveform at the same frequency—and, of course, such a waveform is available in the appropriate timebase.

It is impossible to calculate just how much of each kind of current is required and so preset controls are provided. These are adjusted for best dynamic convergence as seen on a crosshatch display. Basically, there are two presets for each coil, one giving amplitude of parabolic current and the other controlling phase and amplitude of sawtooth current. Left or right displacement (with

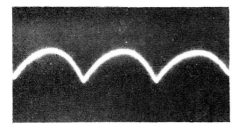

FIG. 6.15. *Parabolic waveforms are present in the field timebase. To these are added suitable proportions of sawtooth waveform to give the correct convergence of the three beams.*

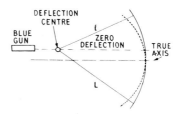

FIG. 6.16. *This diagram shows why a pure parabolic wave needs to be slightly altered for convergence. Here is shown the blue gun only, displaced from the true axis of the tube.*

corresponding tilt of the compounded parabolic current) is governed by the phase of the added sawtooth current, as shown in Fig. 6.18, left displacement (a) and right displacement (b).

Since there are usually three coils for line convergence and three for field convergence (making six coils in three pairs), there may be some twelve presets. In practice, the number may differ depending upon the scope of adjustment provided and on the ideas of the designer.

The blue lateral convergence unit *may* or may not have a winding on it for dynamic convergence—again, depending on the ideas of the designer. Such a unit is always employed, carrying at least the blue lateral *permanent* magnet. If it has a winding, then there is a corresponding preset (or two).

Dual-standard colour sets have an extra convergence problem in that, on changing standard, the dynamic current requirements for line correction are altered (because the components have altered values at the lower frequency).

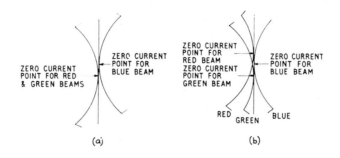

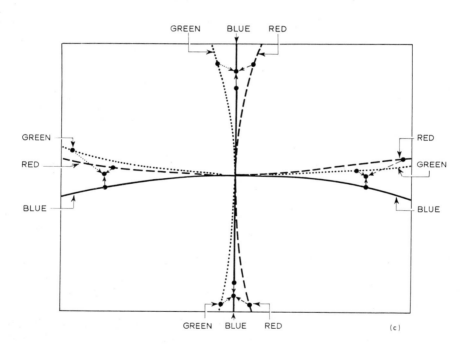

FIG. 6.17. *Correction waveforms (e.g. parabolic modified by sawtooth) required to compensate for the offset positioning of the three guns, (a) field and (b) line. Diagram (c) shows (by the arrows) the dynamic deflecting deviations required to pull the three beams into convergence from the centre upwards and downwards and from the centre to the right and to the left. When these forces have been optimised the three beams trace identical lines (but in three colours, giving white) over the whole of the screen.*

This calls for almost an extra set of dynamic line presets, switched over the two standards, depending on the design.

An associated problem concerns static convergence. This is optimised on one standard but, taking into account the d.c. component of the waveform in the line convergence coils and the resulting fixed magnetic field, it is almost impossible to guarantee that the field resulting from the waveform on the other standard will be the same. This is tantamount to a need for re-adjusting static convergence.

This problem is solved by use of diodes across the convergence coils. These rectify the dynamic waveforms and a series register regulates the d.c. to a value satisfying the static correction needs on the different standard. The fixed

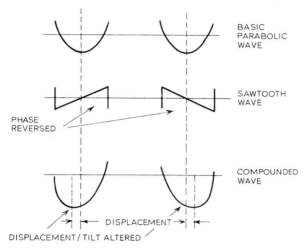

FIG. 6.18. *When sawtooth waveform is added to the basic parabolic waveform, the latter is caused effectively to 'tilt' and displace in a direction governed by the phase of the sawtooth waveform. (a) and (b) show opposite 'tilts' and displacements of the compounded wave when the phase of the sawtooth waveform is reversed.*

field under this condition consists of that provided by the permanent static convergence magnets plus or minus (as required) the field given by the d.c. in dynamic coils. The d.c. can be looked upon as creating a magnetic correcting bias.

A degree of misconvergence is caused by magnetic coupling between the horizontal scanning coils and the convergence assembly, the rear of the scanning coils being close to the convergence assembly. Fig. 6.19 shows what happens. The dotted vertical lines in (a) represent the magnetic field produced by the line scanning coils, while the dots and arrows displaced 120 degrees represent the end-on view of the three beams and their movement in the convergence fields. It will be seen that the line field is vertical where it crosses the blue beam, meaning that deflection of that beam remains in the horizontal direction.

However, the scanning field is collected by the red and green pole-pieces of the shadowmask tube, causing subsidiary deflection of these beams as indicated by the arrows. The result is that when the beams deflect to the left of the screen the green rises and the red falls a little, and conversely when they deflect to the right of centre, as shown in Fig. 6.19.

Modern sets overcome this trouble by the use of symmetry coils across the line scanning coils. These are adjustable by cores which balance out external magnetic fields and thus minimise the coupling to the convergence pole-pieces.

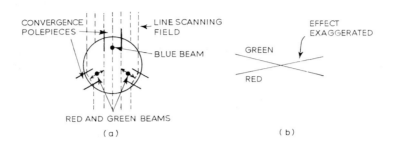

FIG. 6.19. *Influence of line scan field on red/green convergence. (a) shows line scan field in relation to the three beams and (b) the effect on the red and green beams.*

Adjustment is made to remove the red-green tilt with the blue beam off, so that the two lines of a crosshatch pattern combine to form a single yellow line over the whole width of the screen. The circuit involved in this action is considered later. A preset potentiometer may be connected across the field scanning coils for symmetry.

It is worth noting that, if convergence is changed considerably, the tube purity might be affected through the convergence changes shifting the effective position of the guns. This, in turn, could alter the approach angle of the beams to the shadowmask. This is why the convergence should be reasonably correct before finalising the purity adjustment. Some instructions for setting-up bring in a purity check during convergence adjustments.

As we have seen, convergence waveforms are derived from the line and field timebases. The waveforms are fed to a dynamic convergence circuit, usually a printed board or module.

The field section receives two signals from the timebase, a sawtooth voltage and a parabolic voltage, while the line section receives either a line flyback pulse from a separate winding on the line output transformer or a sawtooth current from the circuit of the line scanning coils, depending on the design. Where blue lateral dynamic convergence is adopted, the correction current for this winding is often obtained separately from a line pulse at a tapping on the output transformer, irrespective of the scheme for obtaining the radial correction currents.

There might well be a separate feed and two controls for each of the six radial convergence coils. The basic nature of the two controls is shown in Fig. 6.20. Like the scanning coils in any television set, the field convergence coils are predominantly resistive at field frequency while the line coils are predominantly inductive at the higher line frequencies. This means that voltages are required for the field circuits, while the line flyback pulses can be used for creating saw-tooth *and* parabolic waveform in the line circuits.

Considering the field control first, parabolic voltage is picked up from the cathode or emitter of the field output valve or transistor, and this resembles the waveform shown in Fig. 6.15. This signal voltage is coupled across the amplitude control R1, through C1, so by shifting the slider from chassis side to C1 side the parabolic amplitude is caused to rise.

Sawtooth voltage is fed in at the other end, from R2. This is a centre- 'earthed' preset working in conjunction with a centre-tapped winding on the field output transformer. With the slider at the centre setting shown, there is no sawtooth

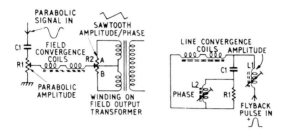

FIG. 6.20. *Basic networks for deriving convergence currents—field (left) and line (right).*

input. Moving towards point A increases the input in one phase, while moving towards B increases it in the opposite phase. It is easy to see, therefore, how the two voltage inputs can be matrixed to give the effects at (*a*) and (*b*) in Fig. 6.18 or any intermediate tilt/displacement as required.

The working of the line control, Fig. 6.20 (right), is more complicated because the convergence coils are mostly inductive. Here a flyback pulse is applied to the circuit and when this meets with inductance a sawtooth current is produced. L1, the amplitude control, is the first inductance encountered. Sawtooth current flows through it and the series resistor R1. It will be seen that the convergence coils are in parallel with R1. Since the inductive reactance of these is much larger than the value of R1, a sawtooth voltage develops across the convergence coils and parabolic current flows in them.

The amplitude of the sawtooth voltage across R1 (and the convergence coils), and the amplitude of the parabolic current in the coils, is adjusted by variable inductor L1. The greater its inductance, the greater the reactance and the smaller the parabolic current in the coils.

In addition, the convergence coils pass a sawtooth current due to some of the sawtooth current in L1 flowing through them in spite of R1 being in parallel. The current of L1 splits, some going through R1 (to give the sawtooth voltage across it), and some through the convergence coils. The sawtooth current in the convergence coils is controlled by C1.

The sawtooth current in R1 is also present in C1 and produces a parabolic voltage across the capacitor. This voltage, in addition to the sawtooth voltage across R1 and the sawtooth current, is applied across the convergence coils.

These signals are matrixed and the net result is a correction current as required for optimum line convergence.

The sawtooth current in R1 is adjustable by L2, in parallel with the resistor. Depending on the inductance of L2, as determined by the setting of its core, so the sawtooth current component is altered. This gives the tilt/phase adjustment.

Practical circuits employ these basic principles. Since the green correction is virtually a mirror-image of the red correction, 'red/green differential' adjustments are often incorporated. In the line circuit this consists of a pair of inductors connected in series with ganged cores arranged so that when the inductance of one coil is increasing that of the other is decreasing. A preset potentiometer does a similar job in the field circuits.

Other designs are focused towards reducing the number of adjustable inductors in the line convergence circuits, since in dual-standard sets some of these have to be doubled-up. The application engineers of Mullard have developed a circuit of this kind, shown basically in Fig. 6.21 (top left).

Here the red, green and blue convergence coils are connected in series with themselves and the line scanning coils. Sawtooth current is fed into the coils from the scanning coils, while a parabolic component is given by the parallel capacitor Cp. Further correction is given by adding a component of current based on the second harmonic. This, in sine-wave form, is generated by the tuned circuit LrCr.

The basic sawtooth current is controlled by the parallel preset RVa, while the modified parabolic current is adjusted by RVd. The R and G correction currents are mirror images of each other, and adjusted in amplitude by RVb. Sufficient integration is given by the inductance of the coils, and parabolic additions are not needed.

The complete dual-standard line convergence circuit using the principles outlined shows the line scanning coils feeding into the convergence circuit through symmetry coils, L6, mentioned earlier. The coils are connected in series with the two halves of the line scanning coils, and the convergence feed is from their junction. By adjusting the cores in L6, the current in the two halves of the scanning coils can be perfectly balanced, thereby removing spurious red/green deflecting fields from the convergence pole-pieces.

It is not particularly difficult to work out the various controls from the description of the basic circuit. It is interesting to note the circuits which are, in fact, changed over on standard changing. The second-harmonic sine-wave tuned circuits are L10 on 405-line standard and L11 on 625-line standard. Their associated presets and the capacitive feed into the circuit from the linearity inductor are also changed. The red/green differential tilt, mentioned earlier, is not changed, and is represented by L2a and L2b.

Diodes D1 and D2 correct the d.c. and thus the static field produced by the convergence coils from standard to standard. Note that the series resistor value, and hence the d.c., is altered by the changeover switch section.

In the Mullard circuit, the blue lateral coils are fed from a tapping on the line output transformer, control being by L12 on 405-line standard and L13 on 625-line standard.

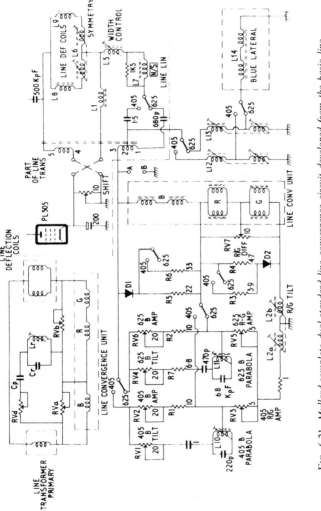

FIG. 6.21. *Mullard complete dual-standard line convergence circuit developed from the basic line convergence system (top left).*

The corresponding Mullard field convergence circuit is given in Fig. 6.22. This uses a simplified system of controls, where, in addition to the red/green differential, a red/green parabolic preset is incorporated.

The sawtooth voltage is picked up from a separate winding on the field output transformer, while the parabolic voltage is taken from a low impedance point in the field output valve cathode circuit. This method of extracting parabolic voltage reduces the amount of degeneration normally encountered by running an effectively low value cathode bypass. Of course, the normal high

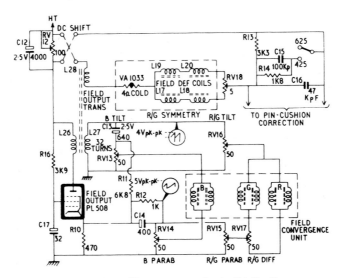

FIG. 6.22. *Field convergence circuit (Mullard).*

capacitance electrolytic bypass cannot be used, but by coupling-in at the 'earthy' side of the capacitor, as in Fig. 6.22, sufficient parabolic voltage is obtained without the amplifier sensitivity being too much impaired.

Field scanning coil symmetry is balanced by RV18, and although the effect on the convergence is less at field frequency, the preset resistor can simplify the convergence at the top and bottom red and green horizontal lines of a crosshatch pattern. These would otherwise tend to run parallel to but separate from each other instead of combining to form yellow lines.

It will be seen in Fig. 6.22 that the convergence coil pairs are connected in parallel (they are in series in the line circuit). This gives a lower impedance and, therefore, improves the matching to the low impedance parabolic voltage feed.

Standard changing does not affect the field convergence circuits, but it is worth noting that single-standard (625-line) colour sets use a somewhat simplified convergence system, and one can find out what this is like by the useful exercise of redrawing Fig. 6.21 omitting the changeover switches and the 405-line standard items.

A complete line and field convergence system, evolved by the engineers of the Thorn-AEI Applications Laboratory, is given in Fig. 6.23. This differs

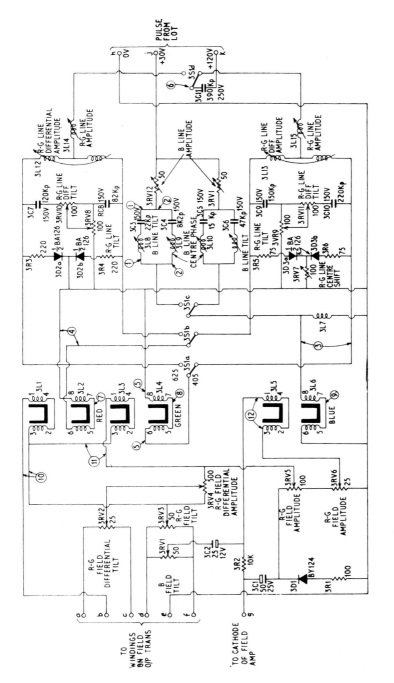

FIG. 6.23. *Line and field convergence circuit (Thorn-AEI).*

85

somewhat from the Mullard circuits as will be understood from the following discussion.

The operation of the line circuit is similar on both standards, and in the 625-line state positive-going line pulses of 120 V amplitude are fed through the red-green amplitude control (3L14) to the centre of the differential amplitude control (3L12). Differential and line tilt are given by networks 3C7, 3RV10, 3C8 and 3RV8. Integrated sawtooth and parabolic voltages are fed to the green coils from 3C7 and 3L12 junction and to the red coils from 3C8 and 3L12 junction. The voltage and current waveforms at these points are illustrated in Fig. 6.24. Line Symmetry (or line balance), dealt with in Chapter 7, is provided by an inductor in series with the line coils.

Static field correction on line changing is provided by diodes 3D2a and 3D2b, in conjunction with their resistors 3R3 and 3R4. These components

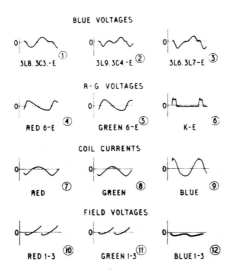

FIG. 6.24. *Complete set of waveforms present in the Thorn-AEI circuit (Fig. 6.23). The numbers here correspond to the number of the test points marked on the circuit.*

also give some shaping to the waveforms. On 405 lines, 3RV7 is used to adjust the d.c., allowing the red/green vertical lines of a crosshatch to be converged without disturbing the 625-line standard setting of the static magnets.

The blue convergence coils are fed with line pulses from 3L8 and 3C3 in parallel with 3L9 and 3C4, with amplitude control by 3RV12. This coil is loaded by 3L7. 3L8 and 3C3 are tuned approximately to the line second harmonic (see the waveforms in Fig. 6.24), in common with the Mullard line circuit.

The basic action of the field side is similar to that already explained. Parabolic voltage is fed from the cathode of the field amplifier and sawtooth voltage from a winding on the field output transformer, via a matrixing network, to the red and green coils. Amplitudes of parabolic and sawtooth signals are adjusted respectively by 3RV5 and 3RV3.

Differential amounts of the signals are controlled by 3RV4 and 3RV2, enabling the spacings of red and green lines to be made equal along the vertical centre line of a crosshatch. Parabolic and sawtooth inputs to the blue coils are controlled respectively by 3RV6 and 3RV1.

Most sets have convergence circuits based on those described. There are differences in detail, but whatever the final configuration—valve or transistor operated—dynamic correction always requires adjustable mixtures of parabolic and sawtooth currents. The dynamic convergence circuit of the BRC 2000 series receivers is given in Fig. 6.26.

HOW TO ADJUST DYNAMIC CONVERGENCE

It has been said that adjusting for optimum convergence is more of an art than a science—this is particularly true of dual-standard sets. It will be discovered that the operation eventually becomes almost instinctive, and adjustments commencing at any point in the sequence will produce the required result. Until such skill is developed, however, the sequence as given by the manufacturer should be followed. This will save time and make for the best results.

It is well worth keeping in mind that absolute convergence is only possible along the centre vertical and horizontal lines of a crosshatch pattern and some degree of edge compromise is necessary, depending on the excellence of circuit design. Shortcomings of the various models in this respect gradually become recognised. This is akin to the variation in linearity performance of monochrome sets.

Before commencing dynamic convergence adjustments, though, it is common practice to ensure adequate degaussing of the tube and its fixtures (avoiding the purity and static convergence magnets). Then the purity magnets are set to zero field (identical tags together). Roughly set the line and field linearity, adjust the focus, adjust the static magnets for the best red/green convergence in a 2 in. circle at the middle of the screen and for the best blue convergence, purify the tube, roughly adjust for dynamic convergence, make final adjustments of linearity, picture size and shift and then commence dynamic convergence according to the maker's instructions.

It will now be instructive to run through a typical purity and convergence exercise of a dual-standard set (BRC 2000 series). Single-standard models are similarly adjusted, but the dynamic convergence adjustments are significantly less involved.

Preliminary Actions

1. Degauss the receiver with external degaussing coil (page 65).
2. Adjust e.h.t. voltage if necessary (page 100).
3. Adjust picture width, height, shift and linearity controls on a test pattern or off-air test card with the colour switched off.
4. Allow the receiver to warm up with the brightness control set to a high level (to warm up the shadowmask) for about 20 min before commencing adjustments.

5. Switch to a 625-line channel.
6. Connect dual-standard pattern generator to aerial socket. (Note: the use of a widely spaced dot or crosshatch pattern is desirable to avoid error of converging on the wrong dots. If a closely spaced pattern is used, however, it is best to make a preliminary coarse adjustment on an off-air picture or test card prior to using the pattern generator.)

STATIC CONVERGENCE ADJUSTMENT

1. Switch off the blue beam.
2. Rotate red and green magnets on the convergence unit to obtain convergence where the vertical and horizontal lines of the crosshatch pattern intersect at the *centre of the screen*.
3. Switch on the blue beam.
4. Rotate the blue magnet on the convergence unit and the tube or assembly on the neck containing the blue lateral magnets to converge blue with red and green at the *centre of the screen*.
5. Switch to a 405-line channel and if necessary adjust the '405 clamp' controls for blue and red/green convergence.

PURITY ADJUSTMENT

1. Switch off the green and blue beams, leaving a *red* raster.
2. Slacken the deflector coil unit clamping nuts.
3. Slide the deflector coil unit backward in its housing to the limit of its travel.
4. Rotate both purity magnets relative to each other to obtain a *uniform red patch at the screen centre*.
5. Slide the deflector coil unit forward until a position is found which gives *an optimum overall red raster*.
6. Lock the deflector coil unit in that position.
7. Check separately on the blue and green beams and repeat the static convergence and purity adjustments until separate red, green and blue rasters with the least contamination from other colours can be obtained.

DYNAMIC CONVERGENCE ADJUSTMENTS

Dynamic convergence adjustments *must follow* the preliminary actions and the static convergence and purity adjustments detailed in the foregoing.

It will generally be necessary to check and re-adjust the static convergence during the dynamic convergence procedure.

The comprehensive dynamic convergence chart relating to the BRC 2000 series receivers is given in Fig. 6.25. This chart shows the location of the various controls on the convergence board which, in common with most receivers, is

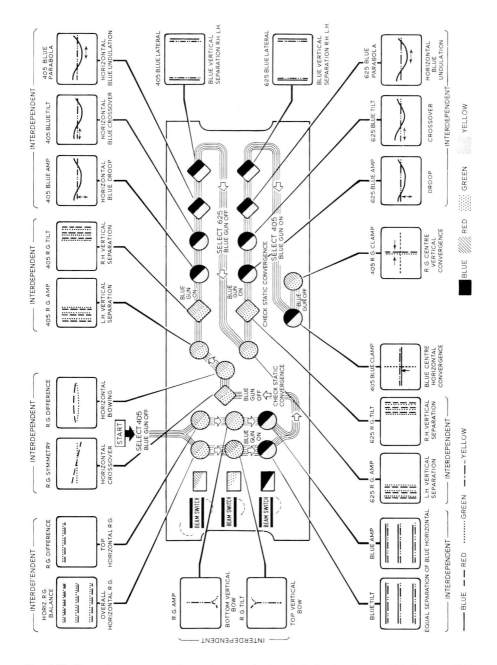

FIG. 6.25. *Dynamic convergence chart relating to the British Radio Corporation (BRC) series 2000 receivers. This shows the location of the various controls on the convergence panel, and identifies their effect on the display (courtesy, British Radio Corporation Ltd.).*

89

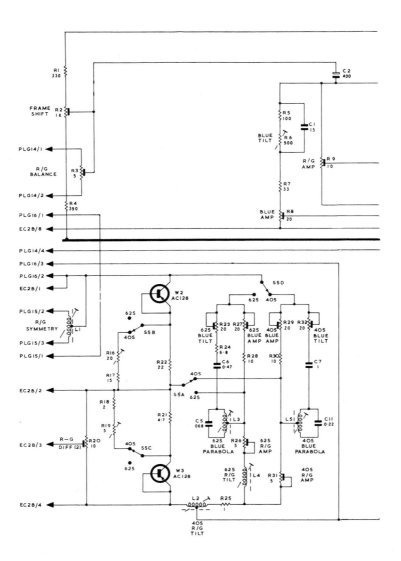

EC2A
1 BOOST HT
2 220V PULSE
3 SOLENOID SWITCH POLE
4 405 SOLENOID
5 625 SOLENOID
6 SPARE

EC2B
1 BLUE CONVERGENCE COIL (LINE)
2 RED/BLUE CONVERGENCE COIL(LINE)
3 RED/GREEN CONVERGENCE COIL (LINE)
4 GREEN CONVERGENCE COIL (LINE)
5 RED/GREEN CONVERGENCE COIL (FRAME)
6 RED/GREEN CONVERGENCE COIL (FRAME)
7 GREEN CONVERGENCE COIL (FRAME)
8 BLUE CONVERGENCE COIL EARTH } FRAME
9 BLUE CONVERGENCE COIL }

EC2C
1 FRAME EARTH
2 FRAME SCAN (IN)
3 GREEN A1
4 RED A1
5 BLUE A1
6 DC EARTH
7 SPARE
8 BRILLIANCE SWITCH POLE
9 BRILLIANCE NORMAL
10 AGC SWITCH POLE
11 FRAME HT FROM POWER SUPPLY
12 FRAME HT TO FRAME BOARD
13 AGC SWITCH
14 30V CHROMINANCE FROM POWER SUPPLY

FIG. 6.26. *Complete convergence circuit of the BRC s*

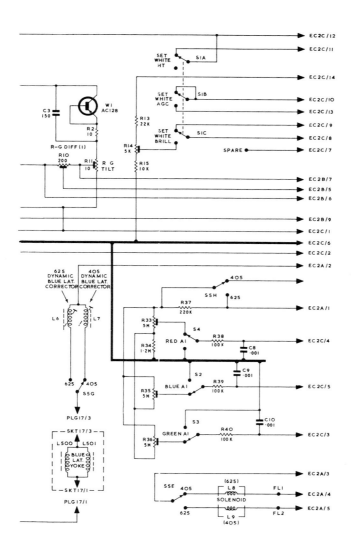

0 receivers (see also Fig. 6.25), showing beam switches.

removed from its normal mounting and placed temporarily into lugs in a vertical position with the controls facing towards the front of the receiver, for ease of adjustment, while the screen is being observed. The convergence board should be carefully restored to its normal mounting, out of the way of inquisitive fingers, after the adjustments have been made.

The chart also identifies the function of the controls, how they affect the crosshatch display, and the sequence of adjustments. It should be noted that

FIG. 6.27. *Convergence board module of the BRC 2000 series receivers (courtesy, British Radio Corporation Ltd.).*

some controls, especially those affecting the same axis, are to some extent interdependent and should, therefore, be adjusted in a complementary manner.

The complete convergence circuit of the BRC receivers in Fig. 6.26 further shows how the variable dynamic convergence components in the circuit influence the three electron beams. The control functions on this tie up with those marked on the chart in Fig. 6.25.

It should be noted that transistors connected as diodes (base and emitter strapped) are used in this circuit to correct the change in standing current through the coils when the line standard is changed. R16 and R19 are the 405 blue and red/green clamps referred to earlier, under static convergence. A photograph of the convergence board is given in Fig. 6.27.

CHAPTER 7

Timebases, E.H.T. and Power Supplies

COLOUR SET TIMEBASES have to supply greater scan power than that required for monochrome sets, as well as the waveforms needed by the convergence circuits. The line timebase in dual-standard sets has to be switchable over 405 and 625 lines, and, therefore, line convergence switching (Chapter 6) is necessary. Moreover, the line timebase has to deliver relatively large e.h.t. power, having in mind three electron beams and the efficiency loss due to the shadowmask. Some sets, however, use a separate e.h.t. generator or voltage trebler.

How well the dynamic convergence remains in adjustment over protracted periods of set operation is governed essentially by the inherent stability of the line and field timebases. This applies to line and field linearity as well as scan amplitude and e.h.t. potential. It is important for the raster geometry to remain as originally set-up for long periods and when changing line standards. It is also important for e.h.t., focus and first-anode potentials to hold steady over normal changes in mains voltage and over full swings of beam modulation.

These factors put large demands on the timebase circuits and, in consequence, they feature the stabilising refinements found in monochrome sets plus others. For instance, the field timebase, as well as the line timebase, may have a kind of a.g.c. system to hold the scan current in the coils steady over changes in supply, load etc. E.h.t. is also stabilised against load changes.

The timebases incorporate electromagnetic picture shift controls which take the place of the shift magnets of monochrome sets. Then there is commonly (but not always) a rather special arrangement for counteracting residual pincushion distortion.

FIELD TIMEBASE

A Mullard field timebase circuit for colour sets is shown in Fig. 7.1. This provides scan power for a 90 degree tube along with convergence waveforms. The complete field convergence circuit which ties up with this timebase was given in Chapter 6.

The circuit is relatively simple, using $\frac{1}{2}$ECC82 triode in conjunction with a PL508 in a self-oscillating arrangement (multivibrator). The time constant, hence field frequency, is given by the RC elements in the grid circuit of the triode, the former including a potentiometer for vertical hold control. Feedback is from the anode of the PL508 to the grid of the ECC82 via a capacitor and resistor and there is the usual forward coupling from the anode of the triode to the control grid of the PL508. Field linearity is ensured by a feedback network from the anode to the grid of the PL508 following normal practice. Two presets in this network allow for adjustment of overall linearity and linearity at the top of the raster.

The OA81 diode is in a field sync gating circuit. The front of this circuit receives sync signals from the anode of the sync separator valve, and the output

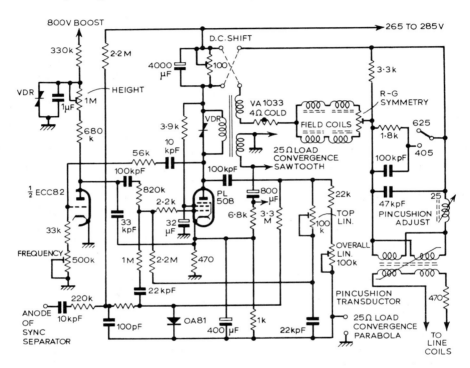

FIG. 7.1. *Field timebase for colour sets using 90 degree tube, also providing convergence currents (Mullard).*

supplies sync pulses to the grid of the PL508. The gating action is such that the diode opens during the retrace and closes during the scanning stroke, so preventing line interference reaching the PL508 grid. This gives good interlace performance.

To help achieve very low scan non-linearity (about 3 per cent) the triode is fed from the boost h.t. supply (about 800 V) and stabilisation is given by the v.d.r. (voltage-dependent resistor) forming part of the potential divider in the height control circuit. If the supply voltage falls, the value of this resistor rises and, since it forms the bottom leg of the potential divider with the 330 kΩ resistor forming the top leg, the voltage applied to the height control remains substantially constant. Stabilisation is also assisted by the boost voltage being derived from the stabilised line timebase.

The v.d.r. across the primary of the field output transformer has the effect of muting the field retrace pulses. During the scan stroke the voltage build-up across the transformer is relatively low and the v.d.r. is at high value. On the high pulse voltage retrace, its value falls, thereby reducing the amplitude of the pulses. Even so, the field retrace pulse amplitude is generally above that in monochrome counterparts, which is a good reason for keeping the fingers clear of the field output valve anode circuit.

VERTICAL SHIFT CONTROL

The circuits so far described will not be strange to the service technician, but we now come to those sections peculiar to colour set timebases. The vertical shift control is one such. Fig. 7.1 shows that the field output valve current is passed through the field scan coils but limited by the shunt-connected shift control and the series resistors. This arrangement provides a small d.c. in the coils without undue interaction on the field scan current delivered by the transformer.

The amount of d.c. in the coils is adjusted by the shift control. This simply varies the d.c. potential developed in series with the 'dead' side of the field output valve anode. With the resistance at zero there is no d.c. flow in the shunt path, while at maximum resistance setting there is current in the parallel scan coils. The field frequency itself is by-passed across the resistive element by the 4000 μF electrolytic capacitor. It is possible to reverse the d.c. in the scan coils to get the picture to shift more in the opposite direction. This reversing device is generally nothing more than a pair of soldered connections, represented by the dot lines.

The field output transformer carries two secondary windings. One is for energising the scan coils via the shift control. The other delivers a sawtooth waveform to the field convergence circuit.

Fig. 7.1 also shows the red-green symmetry control which simplifies the convergence of the top and bottom red and green *horizontal* lines of a cross-hatch pattern, which would otherwise tend to run parallel but separate, instead of forming single yellow lines (blue gun out). This is fully explained in Chapter 6.

The parabolic waveform for the field convergence coils is taken from a low impedance section of the output valve cathode circuit. This required waveform is present at the cathode of the field output valve but, to obtain adequate amplitude, a part of the cathode circuit needs to be unbypassed, which causes degeneration and impairs the sensitivity of the amplifier. This is combated in Fig. 7.1 by extracting the signal from the bottom end of the cathode circuit via a 25Ω load (given by the convergence circuit itself).

PIN-CUSHION, CAUSE AND CORRECTION

Before we can appreciate the pin-cushion correcting circuit, it is necessary to have a better understanding of this type of distortion, its cause and cure. Fig.

7.2 illustrates exaggerated pin-cushion distortion. This results from wide-angle scanning and a relatively flat screen and, in monochrome sets, is corrected by a pair of small magnets near the tube flare. These 'pull out' the curves at the edges. Permanent magnet correction is impossible on a colour tube using three beams, so arrangements are made to 'modulate' the vertical scan current with a current coupled-in from the line circuits, and vice versa, to correct for residual pin-cushion distortion remaining after dynamic convergence. A special 'transductor' component is used for this purpose (see Fig. 7.1).

How scan correction is achieved by this cross coupling between timebases can be appreciated by studying Fig. 7.2. The sides of the raster are straightened by making the *horizontal* deflection greater in the centre of the vertical scan than at top and bottom. The transductor handles this by providing maximum amplitude at the centre, reducing progressively to zero correction at the top and bottom.

Top and bottom edges of the raster are straightened by applying correction to the *vertical* scan at line frequency. Take the top scan line, Fig. 7.2. When this line is being traced on the screen, the correction is zero at the start of the line, increases to maximum at the centre and then decreases to zero (on the opposite phase) for the second half of the line.

It is evident from Fig. 7.2 that the required correction diminishes from line to line, down the field, to zero at the centre line and then rises again to maximum correction at the bottom line. The correction per line is provided by a parabolic current superimposed on the field current waveform, as shown in Fig. 7.3. It is impossible to scale this diagram, since it would be necessary to draw $312\frac{1}{2}$

FIG. 7.2. *Pin-cushion distortion of the raster shown in exaggerated form.*

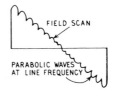

FIG. 7.3. *Waveform required to correct pin-cushion distortion (not to scale).*

parabolic waves (625 lines) on the one field scan wave. However, the idea is shown along with the reducing amplitude waveforms and the necessary phase reversal from the centre downward.

In Fig. 7.1 the amount of pin-cushion correction fed into the field scan coil circuit from the transductor can be adjusted by the variable inductor L25. For optimum working on both standards, it is necessary to modify the transductor coupling slightly on the 405 standard and this is done by the line-change switch section bringing in the 1·8 kΩ resistor and the 100 kpF capacitor.

Finally, to combat the change in the load on the field amplifier, which could result from the resistance of the field scan coils rising with increase of temperature (with consequent fall in vertical scan amplitude), a thermistor (VA1033) is included in the scan coil assembly. The resistance of this falls with increase in temperature, countering the rise in scan coil resistance.

A slightly different circuit, using low-impedance scan coils, is shown in Fig. 7.4. This uses a pair of triodes in a multivibrator driving the pentode section of a PCL85 field output. Apart from the generator this circuit has much in common with the previous one especially in terms of vertical shift and stabilisation. The

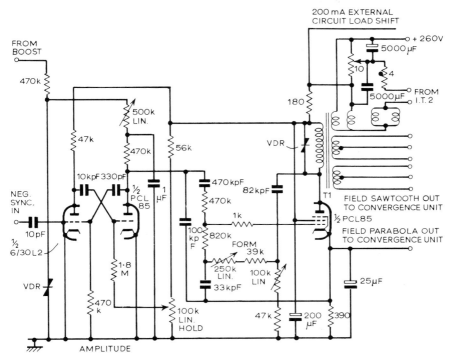

FIG. 7.4. *Field timebase employing triode multivibrator and pentode output to feed low impedance scan coils. Vertical shift and stabilisation are similar to the methods adopted in Fig. 7.1.*

convergence coils are driven differently, however, and a pair of windings for this purpose are present on the field output transformer. These are connected to the Thorn-AEI convergence system detailed in Chapter 6. The parabolic output is taken direct from the cathode of the field output valve, as in the Mullard circuit.

A part of the field timebase circuit from the Philips G6 colour chassis (covering type G25K500 models) is shown in Fig. 7.5. This is an interesting circuit in that it employs a triode 'buffer' stage between the two-triode multivibrator (not shown, but fairly conventional) and the field output stage and also field stabilisation.

FIELD BUFFER STAGE

The buffer triode is connected as a cathode-follower having a high input impedance and thus presenting little load on the generator and a low output impedance for power coupling to the output valve (PL508).

Line-timebase-type stabilisation is provided by the d.c. voltage developed in the control grid circuit of the PL508, due to the v.d.r. rectifying pulses fed back via C1. The d.c. is negative and partly countered by a positive potential from the 'field stabilisation' preset, which picks up its supply from boosted h.t. rail. The preset alters the grid bias. Adjustment is for optimum operating

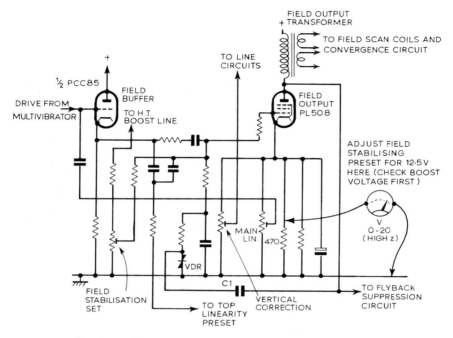

FIG. 7.5. *Field timebase using a triode 'buffer' (Philips G6 series)*.

conditions, which is obtained when 12·5 V is registered across the cathode resistor. If the field drive decreases or the load increases for any reason, the valve turns 'on harder' owing to the negative bias falling.

Because the field preset is fed from the boosted h.t. rail, the line stabilisation must be adjusted first and, in the Philips sets, this is done to obtain a boost of 560 V on 405 and 600 V on 625.

The 'vertical correction' preset feeds field signal from the cathode to the line circuits. Adjustment is made on a crosshatch display for straight vertical lines at the screen centre. Pin-cushion is corrected by the usual transductor and variable inductor for straight horizontal lines at top and bottom.

Fig. 7.6 shows the line output stage associated with the Mullard line convergence circuit, described in Chapter 6. The PL505 line output valve is heftier than the monochrome counterpart because, apart from extra scan power and the convergence signals, the line output stage has to deliver e.h.t. in excess of 1 mA at 25 kV. (The Mullard colour valves are partnered by the PY500 booster diode, the PD500 shunt-stabiliser triode and the GY501 e.h.t. rectifier.)

The circuit uses low impedance convergence coils in series with the line scan coils, and the line output transformer uses the desaturation technique, where windings carry d.c. of opposing polarities. This neutralises the field to allow greater magnetic swing for the signals. Harmonic tuning, switched over both standards, is also used. Extra windings on the transformer produce convergence signals, while the blue lateral convergence signal is delivered at the 405-line tapping, the pulses here being about 200 V amplitude.

STABILISATION

Harmonic 625-line tuning is given by L4 and for 405 operation L5 is switched in series. The e.h.t. winding is between tags 12/13 and the output is applied to the e.h.t. rectifier anode in the usual manner. It is essential for the e.h.t. to hold

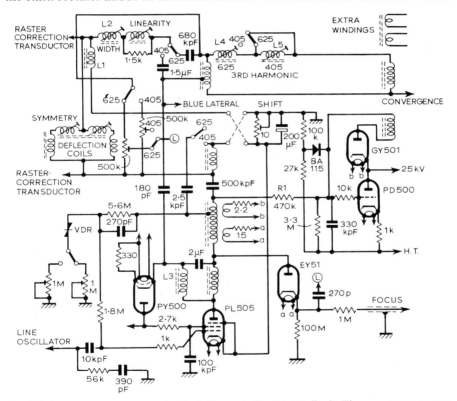

FIG. 7.6. *Line output stage, partnering the field stage in Fig. 7.1 (Mullard). The convergence arrangements associated with this are detailed in Chapter 6.*

constant over full swings of beam current to ensure faithful colour reproduction at all brightness levels. This is achieved by the stabiliser triode PD500 connected (anode-to-cathode) across the e.h.t. supply. How much e.h.t. current it passes is controlled by the voltage on its grid.

In the circuit the conventional overwind feeds the anode of the e.h.t. rectifier GY501, and in shunt with the supply is the PD500 stabiliser. The current passed by this valve is controlled by its grid voltage, and the voltage is obtained from a divider circuit which samples the e.h.t. current from the overwind. The 'cold' end of the overwind is returned to the boosted h.t. supply through R1 (470 k) and is also connected to the h.t. line through the 3·3 M resistor. The junction is connected to PD500 grid through the 10 k resistor. The flow of e.h.t. from the 'cold' end of the winding makes the junction and hence the grid negative of a value dependent on the current flow. Thus an increase in beam

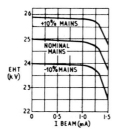

FIG. 7.7. *Performance of the shunt stabiliser in Fig. 7.6.*

current results in a negative voltage increase at the PD500 grid which turns down the current shunted by the valve, thereby making more available to meet the increase beam current of the tube. The performance of the stabiliser is shown in Fig. 7.7.

Diode BA115 is a 'monitoring diode' for beam current limiting. This is necessary because the stabiliser can do nothing to control a fault resulting in excessive current. Should the diode detect excessive e.h.t. current through the overwind conduction occurs and a 'turn-down' bias is reflected to the brightness control circuit.

FOCUS AND E.H.T. ADJUSTMENTS

The focus electrodes of the shadowmask tube require a potential of about 7 kV and this is obtained from the separate EY51 (or equivalent) rectifier processing pulses of suitable amplitude at a tap on the line output transformer. Fine adjustment is available on both standards by the focus presets. These work by varying the amplitude of the pulse fed to the 270 pF reservoir capacitor, and this method of adjustment keeps the potential between the preset spindles and the chassis relatively low.

Horizontal shift is given by the voltage developed across a 10Ω preset connected in the cathode of the line output valve. This is fed through the line coils by the 'hold off' indicator L1 and, as with the field shift, a means of reversing the current in the coils is provided, shown by the dot lines across the shift preset.

Some sets have a separate preset adjustment for e.h.t. The set-up in the Philips G6 chassis is shown in Fig. 7.8. The e.h.t. is set by connecting a high-impedance voltmeter across the 1 kΩ cathode resistor and, with the tube cut-off (by depressing the luminance switch), adjusting the preset for a reading of 1·2 V. This means that the stabiliser is then passing 1·2 mA.

The boost and e.h.t. adjustments are very important since maladjustment can cause the e.h.t. to rise above the nominal 25 kV. While this may not particularly harm the tube, it will certainly result in a rise of x-ray emission, possibly forward from the tube face.

The e.h.t. rectifier and stabiliser triode can emit x-rays in excess of the permissible dose rate, even under normal conditions. Heavy shielding surrounds these parts of the set, and it is most important for the shielding to be correctly refitted after servicing. It would be dangerous to run the e.h.t. circuits with the

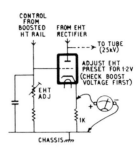

FIG. 7.8. *Method for setting the e.h.t. voltage (Philips G6 series).*

screens removed while working close to these valves. Some sets have interlocking switches, killing the e.h.t. when the screens are removed.

The line timebase in Fig. 7.9 is a development of Thorn-AEI and ties in with the Thorn-AEI convergence system described in Chapter 6. A simple two-triode line multivibrator is employed, which is switchable over the two line speeds. To provide the required scan and e.h.t. power a pair of monochrome-type line output valves and booster diodes are used in this circuit, and the design ensures that satisfactory performance is maintained with low limit line output valves near the end of their operating life.

Stabilisation is by v.d.r. and by the 6BK48 shunt-stabiliser triode. Control is from the boosted h.t. line—an increase in tube current causes the voltage to fall and this is communicated back to the grid of the triode causing its current to drop. The v.d.r. in series with the boost rail virtually amplifies the change in boost voltage, making the control more effective.

With the tube cut-off, the v.d.r. presets are adjusted to establish the correct boost given by a stabiliser current of 1·25 mA in the cathode resistor. The e.h.t. preset is then adjusted to yield an e.h.t. of 25 kV.

Horizontal shift in this circuit is provided by the voltage developed across a resistor passing a portion of the mean booster diode current. Reversal facilities are also provided as in the other circuits. Focus potential is derived by the use of a selenium rectifier receiving pulses from a tap on the line output transformer, and adjustment is provided by a 10 MΩ preset.

The transductor system of pin-cushion correction is adopted and the component is I.T.2 in the circuit.

The field timebase of the BRC 2000 series receivers is given in Fig. 7.10. This

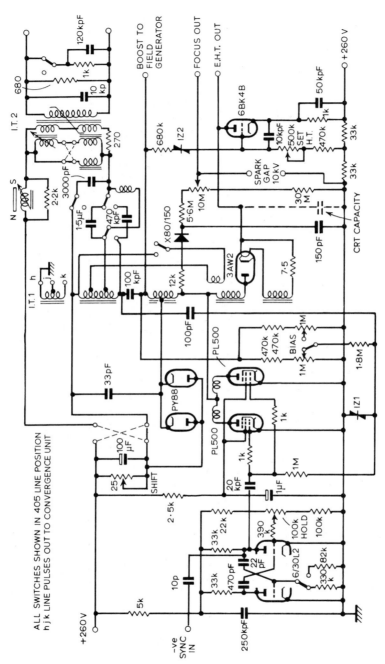

Fig. 7.9. *Line oscillator, output and e.h.t. circuit. The necessary power is provided by the PL500 and the PY88 arranged in pairs (Thorn-AEI).*

102

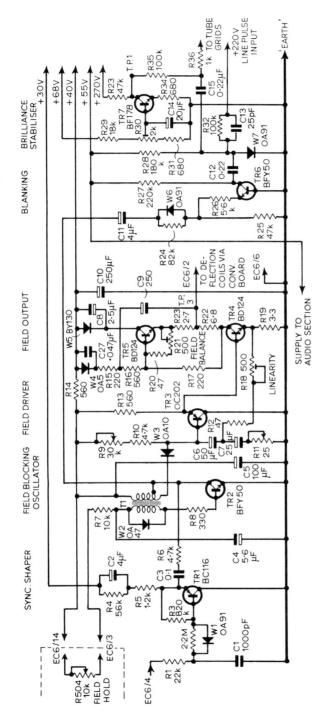

Fig. 7.10. *Transistorised field timebase (BRC).*

103

uses a fairly conventional blocking oscillator (TR2), with feedback via transformer T1 from the collector to base. The field hold control (R504) adjusts the time-constant, and hence the repetition frequency.

The switching signal, picked up from T1, is fed to the field driver common-collector stage (TR3), via diode W3, giving a low impedance drive to the base of the field output transistor (TR4). This works in conjunction with its push-pull partner (TR5), receiving the drive signal at its base, via R17, from TR4 collector. Feedback is applied to TR5 through R20 and R21 and, since the latter is a preset component, the conduction of TR5 can be balanced with that of TR4. The two transistors (TR4 and TR5) thus switch alternately, and field current is fed to the scan coils from the emitters, via the convergence board (see Chapter 7). Overall field linearity is set by R18, which adjusts the feedback from the output stage to the base of the driver transistor, via a frequency-selective network comprising C6, C7 and R11. The last is a preset which influences the linearity at the extreme of the scan.

The field balance preset sets the output stage crossover point, and the procedure for adjusting this type of circuit is as follows:

1. Adjust the height and linearity.
2. Re-adjust height.
3. Adjust the balance preset for a reading of 24 V at the emitter feed to the field scan coils (Test Point 3—T.P.3—on the circuit.)
4. Re-adjust height.

Incorrect crossover will show as field distortion (e.g. as a horizontal band on a raster).

The sync shaper stage (TR1) receives field sync pulses at its base and applies them at the correct impedance amplitude and polarity to the collector of the blocking oscillator transistor (TR2) for optimum interlace and field lock.

Although the blanking and brilliance stabiliser stages (TR6 and TR7) are not directly concerned with the field circuits, they are located on the field module. Field pulses are fed to TR6 emitter through C11 and the diode/resistor combination (W6 and R24), and appear negative-going at the collector, since the stage is in common-base configuration. These pulses, along with line blanking pulses, applied through R32 and C13, are then fed to the grids of the colour tube through C15 and R36, blanking the guns on the retraces in the usual way.

The grids of the picture tubes receive a standing bias potential from the 270 V line through R23 and R35, and as R23 with TR7 act as a potential divider in the circuit to the grids, the potential is affected by the degree of conduction of TR7. The conduction is set by the base potential divider (R29, R30 and R31), and since R30 is a preset a biasing adjustment is provided. The bias potential is also, in this way, stabilised against changes in supply voltage. The correct biasing potential is established at Test Point 1 (T.P.1) at TR7 collector, by adjusting R30.

The field timebase module of the BRC 2000 series is shown in Fig. 7.11. This also accommodates the audio section, using six transistors, with the two drivers and the two outputs arranged in complementary push-pull (see Chapter 9).

While many colour sets use transistors for the low power stages, valves are still employed for the higher power stages, including the line timebase and e.h.t. sections. All-transistor models, of course, use transistors and diodes in all

FIG. 7.11. *Field timebase module (BRC 2000 series). This includes the audio, sync shaper, blanking and brilliance stabiliser sections (courtesy, British Radio Corporation Ltd.).*

sections, including the line timebase and the e.h.t. The trend is towards the use of an e.h.t. trebler circuit, driven from the line timebase, and the following is an account of the design aspects of one such arrangement.

DESIGN ASPECTS OF ALL-TRANSISTOR LINE TIMEBASE AND E.H.T. CIRCUITS

A conventional black and white receiver requires, typically, about 3 W of screen power to provide a picture of reasonable brilliance. Shadowmask tubes, as we have seen, only permit a small percentage of the total energy to reach the screen phosphors. Some 30 W of beam power is required to provide an acceptable picture.

It is normal for the e.h.t. to be produced from the high flyback voltages developed in the line scan transformer, as already explained. However, dual-standard working complicates the process because the energy and voltage available to e.h.t. is variable with the width settings and different frequencies. The flyback time is fixed by the transmission standards at approximately 12 μs on 625 lines and 18 μs on 405 lines; this can be embarrassing.

Fig. 7.12 shows a basic transistorised line scan circuit in which the output transistor (there are two in the real circuit) is driven by square wave. When the drive is positive-going, the transistor is switched on and current flows from the 60 V line through the inductance of the scan coils and transformer, and rises in a slow exponential to produce sawtooth section A–B.

At point B the transistor is switched off, and then T1, the scan coils and C1 form a resonant circuit so the current falls at the natural frequency of this combination (flyback time). This rapid fall of current induces a high voltage across the transformer and deflector coils. It is this high voltage peak which is used in a conventional flyback e.h.t. system to produce the required voltage by means of a transformer.

To achieve an all-transistor receiver, the e.h.t. rectifier needs to be a selenium or similar type which can handle the larger currents and voltages when used

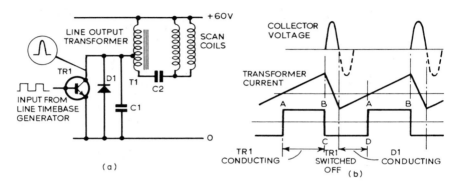

FIG. 7.12. *Basic line scan circuit in which the output transistor is driven by a square wave (a). Details of the operation of the circuit are shown at (b).*

in a trebler circuit. A practical compromise has a forward resistance which is defined by a forward current of 14 mA and 1 V per rectifier plate applied; i.e. $70\,\Omega$ per plate. Using the trebler arrangement, the three forward pulse conducting rectifiers are significant, as each one is assembled with 200 plates, the effective forward resistance is $200 \times 70 \times 3 = 42\ \mathrm{k}\Omega$. If the rectifiers were conducting continually this would be the true forward regulation resistance. However, the flyback pulse represents only a small fraction of the operational time cycle. Fig. 7.13 illustrates an idealised pulse, and shows that in practice a rectifier with a forward resistance of $42\ \mathrm{k}\Omega$ is in fact effectively nearer to $4\cdot2\ \mathrm{M}\Omega$, assuming a conduction period 1/100th of the total cycle. This means that to obtain 24 kV at 1 mA, $28\cdot2$ kV is required to allow for the rectifier drop, shown by the diagrams at (*a*). If the pulse duration time is doubled, and consequently the conduction time, then the regulation is improved pro rata and only $26\cdot1$ kV is required for the same output voltage, as shown at (*b*). It is thus advantageous to have the longest practical pulse duration, or to increase the repetition rate. Either will improve the mark/space ratio, and hence the regulation.

When e.h.t. is derived from the line scan transformer, neither the flyback time nor the repetition rate is flexible due to transmission standards. It is thus logical to separate the e.h.t. generation from the line scan transformer. This is done in the BRC all-transistor sets.

Fig. 7.14 illustrates the separate e.h.t. generator. This is almost identical to the line scan stage except that there are no scan coils. It is separate from the

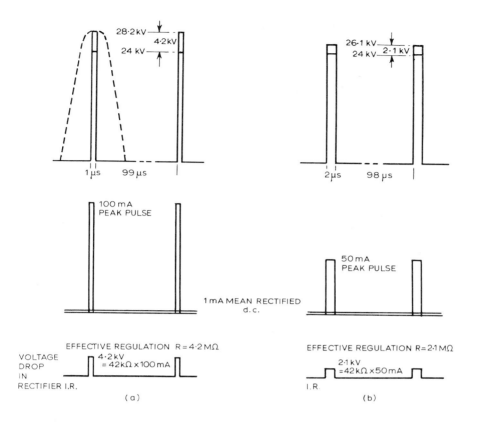

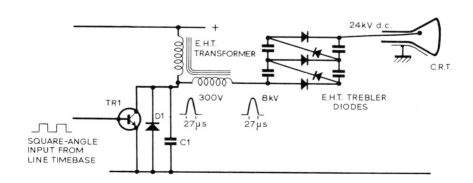

FIG. 7.13. *Pulse conduction of the e.h.t. rectifier. (a) effective regulation when the source equals 4·2 megohms and (b) when it equals 2·1 megohms, as obtained by doubling the pulse duration time.*

FIG. 7.14. *Basic e.h.t. generator with five rectifiers in a voltage-trebler (BRC 2000 series).*

107

scan stage with the exception of the drive waveform which is derived from the common line source. While this controls the repetition frequency, the flyback time is separately controllable and is lengthened to approximately 27 µs. Although the regulation is thus improved, it is still not good enough, and further improvements are necessary.

There are two solutions available; one is to have a compensatory shunt load, but this is very difficult with solid-state devices. The second is by compensating for variations in output by changing the voltage input by means of a series regulator, which is the solution adopted by BRC.

Although a series regulator sounds relatively simple, it is in fact the most complex part of the e.h.t. circuitry. Fig. 7.15 illustrates the basic circuit. Feedback from the e.h.t. output is sampled at the input and the output is controlled to compensate. The sensitivity is such that a change at the input of 150 µW will

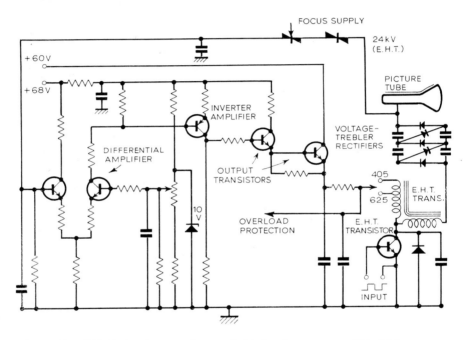

FIG. 7.15. *Basic voltage regulator used with the e.h.t. generator of Fig. 7.14.*

cause an output change of 30 W. This is the typical operating condition, and the response is virtually flat from zero to 1 kHz. With this performance characteristic the regulation resistance is reduced from 6 MΩ on 625 lines and 10 MΩ on 405 lines to a common regulation of better than 1 MΩ, which is a satisfactory level.

The regulator in Fig. 7.15 behaves admirably under controlled conditions, but it provides no safeguard against overload, and if the brightness is increased the circuit will continue to supply power into the output stage until it eventually destroys itself. The development of the circuit in Fig. 7.16 incorporates a 'sensing' transistor, which samples the mean current of the e.h.t. generator, via the negative line, and reduces the range of the brightness control in sympathy.

The zener diode samples the voltage drop across the pair of output transistors and prevents forward currents in excess of 1 A developing. It also makes the forward rise-time of the regulator commensurate with the fall-time, thereby stabilising the regulator.

The complete line timebase circuit is given in Fig. 7.17. This is in three main sections, which are: line oscillator (VT1 and VT2), line driver (TR3) and line

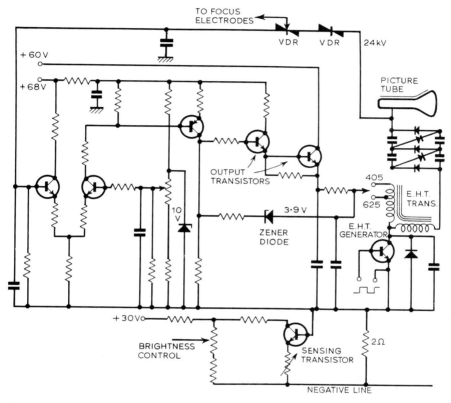

FIG. 7.16. *Development of the circuit in Fig. 7.15, showing the 'sensing' transistor and the zener diode connected across the output transistors. The action of these components is explained in the text.*

output (TR4 and TR5). Diodes W1 and W2 are arranged in the form of a 'phase detector' circuit, which produces a control potential at the base of TR1 when the phase (and hence frequency) of the sample line pulses fed in through R2 drifts from that of the line sync pulses (fed in at S500—the line sync shorting switch on the 'push and twiddle' control).

The control potential corrects the phase (and frequency) of the line 'signal' to match the sync pulses, in the conventional flywheel-controlled line sync manner. The line generator is a form of blocking oscillator, in which L2 and L3 give the 405-line timing while the 625-line timing is achieved when L1 is switched in parallel with L2 by S1D.

LINE OSCILLATOR

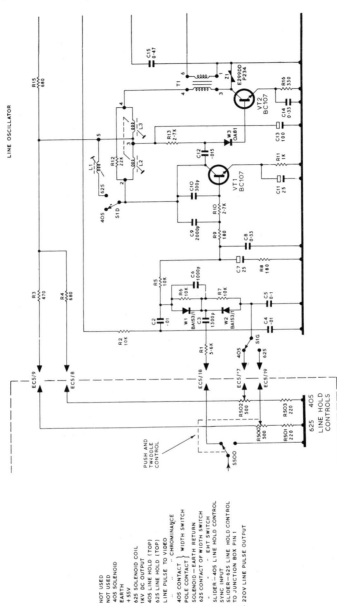

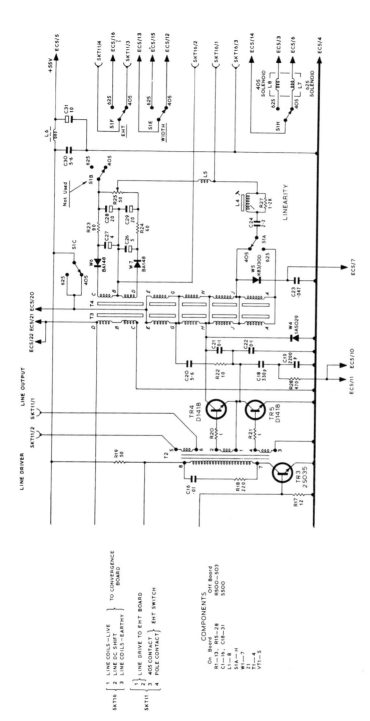

Fig. 7.17. *Transistorised line timebase (BRC 2000 series).*

Manual control of frequency is provided separately by the 405-line and 625-line hold controls, selected by S1G. These are potentiometers across the supply, which give a control potential from their sliders in series with that obtained from the 'phase detector' diodes.

TR1 circuit can be considered as a variable electronic reactance arranged to adjust the generator frequency under the control of the 'phase detector' diodes.

The oscillator is coupled to the driver by transformer T1, while the driver is coupled to the two output transistors by transformer T2. The design of the driver and its transformer is such that square-wave pulses, with sharp leading edges, are produced to switch the output transistors rapidly between their on and off conditions. This is necessary because of the limited power handling capacity of the output transistors.

The sawtooth current required for scanning is created from the square waves partly by the utilisation of the non-linear characteristics of the transistors and partly by the action of the inductance of the line scan coils and the output transformer, which is in two parallel-connected 'jellypot' sections.

Diodes W6 and W7 provide a d.c. voltage between the slider of the horizontal shift control, R25, and point B on the line output transformer section T4, with the associated electrolytic capacitors acting as 'smoothers'. This is for operating the horizontal shift circuit. Diode W5 rectifies pulse voltage at point J on the same transformer and delivers 1 kV d.c. across the reservoir capacitor C23.

Line linearity works in the conventional way, via the inductor L4. While the feeds and outputs are identified on the circuit, it should be mentioned here that the standard-change switch sections associated with the various modules are operated by solenoids, which themselves are energised together from a primary 'systems' switch and circuit.

The square-wave drive for the e.h.t. generator is derived from a winding on the secondary of T2 (points 5 and 6), and this is fed to the e.h.t. generator transistor (TR7) in the e.h.t. circuit, given in Fig. 7.18. The development of this circuit has already been explained, but from the practical operating viewpoint, the regulator output transistors (TR5 and TR6 in Fig. 7.18) are arranged automatically to adjust the direct supply current fed to the e.h.t. generator transistor (TR7) from the +58 V line (EC4/1), via the primary of the e.h.t. transformer T1.

The current reaching the e.h.t. transistor is thus governed by the degree of conduction of TR5 and TR6, which in turn depends on the control voltage at the base of TR5. This control voltage is derived from a sample of the e.h.t. voltage picked up from the e.h.t. line, via two voltage-dependent resistors (in series), the differential amplifier (TR2 and TR3) and the inverter amplifier (TR4). Thus, should the e.h.t. voltage fall, greater power is delivered by TR5 and TR6 to the e.h.t. transistor, resulting in a compensating rise in e.h.t. voltage. This is the same effect as the e.h.t. supply source resistance being decreased, as explained in the design notes. Regulation characteristics to suit the line standard are switched by S1F, supplying the d.c. input to point A on the e.h.t. transformer on the 405-line system and to point B on the 625-line system.

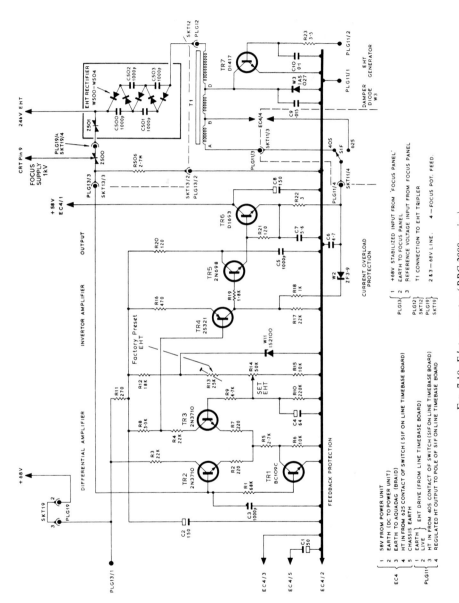

FIG. 7.18. *E.h.t. generator (BRC 2000 series).*

113

(a)

(b)

FIG. 7.19. *Line timebase module (a) and e.h.t. generator (b). Note the two 'jellypot' line transformers in (a) (BRC 2000 series) (courtesy, British Radio Corporation Ltd.).*

114

Refinements include feedback protection (TR1) and current overload protection (W2). The e.h.t. voltage is set by adjustment to the gain of the differential amplifier by the potentiometer R14 (with a factory preset adjustment R13 in series with it). The correct setting is best established by the use of an e.h.t. voltmeter, for a reading of 24 kV.

SPARK GAP PROTECTION

To protect the solid state devices of contemporary sets from damage which could result from random but natural discharges across the electrodes in the picture tube, a number of spark gaps are fitted round the tube base as shown in Fig. 7.20(a). Sometimes the gaps are formed between the ends of wire electrodes

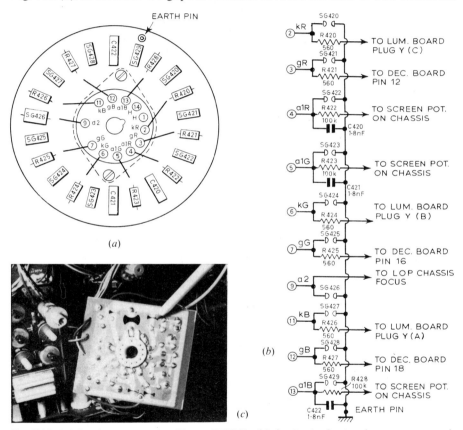

Fig. 7.20. *(a) Spark-gap protection (Decca CTV25). (b) the circuit, showing the components on the tube base tags. (c) Printed circuit spark gaps of the BRC 3000 series, single-standard models.*

connected to the tube base tags and to an 'earthing' ring or plate, but spark gaps of printed circuit form or component type may be used. The circuit arrangement used in the Decca CTV25 colour set is shown in Fig. 7.20(b), which also reveals the various resistors and capacitors, which assist with the protection, connected to the base tags, and the points in the set to which they are terminated.

The gaps are chosen or adjusted so that the energy of any discharge in the tube is bypassed from the circuits connected to it, via the spark across the gap and a low resistance cable to the outer conductive coating of the picture tube, connected to 'chassis'. Without such protection the short-duration discharge energy, which can be as high as 700 amperes, can reflect sympathetic discharges across the components connected to the tube electrodes, and high-peak transitory-type voltages can be reflected from circuit to circuit, where they will almost certainly find their way into the junctions of semiconductor diodes and transistors, resulting in their immediate destruction. The series resistors increase the impedance of the external circuits so far as these untamed discharge currents and voltages are concerned and thus make the bypassing more efficient. Capacitors may also be used in the circuit. Fig. 7.20(c) shows the printed circuit spark gaps in the BRC 3000 series sets. The associated components, on the other side of the circuit board, must not be shorted or varied in value.

POWER SUPPLIES

Hybrid sets need supplies at ordinary h.t. voltages for the valves and lower voltages for the transistors. These are not difficult to obtain, as shown by the Decca CTV25 power supply circuit in Fig. 7.21.

A single mains transformer is used with the conventional h.t. primary acting as an auto-transformer step-up on all inputs other than 240 V. A single half-wave h.t. rectifier, fed through an 8Ω surge limiting resistor, delivers the d.c. output, choke/capacitor smoothed in the usual manner, into three separate feeds, each filtered by a resistor and electrolytic capacitor.

Three transistor supplies, at -18 V, $+18$ V and $+25$ V, are obtained from a separate secondary winding and two half-wave rectifiers, with appropriate filtering and smoothing. The 6·3 V supply for the heaters of the colour tube and a pilot light is obtained from an l.t. winding. It is not uncommon to find the heaters of the valves connected in series, and energised direct from the mains supply, via a mains dropper, in the a.c./d.c. set manner. However, colour sets must not be connected to d.c. mains supplies because of the mains transformer.

Protection is given by a 2 A mains input fuse and a 1 A h.t. supply fuse. It should be noted that this type of power supply gives the set a mains-connected 'earth' line, owing to the primary acting as an auto-transformer.

STABILISED AND REGULATED SUPPLIES FOR ALL-TRANSISTOR SETS

The power supply system for the all-transistor BRC models is in two parts, comprising the power supply proper and the power supply regulator. The circuits of these are given respectively in Figs. 7.22 and 7.23 (pages 118-120).

These rather elaborate arrangements are desirable with transistorised equipment to maintain constant supply voltages in spite of relatively wide variations in load current. The two circuits are complementary and should be studied together.

Zener diodes and transistors are employed for stabilisation and regulation, and the various supplies are obtained from three isolated secondary windings

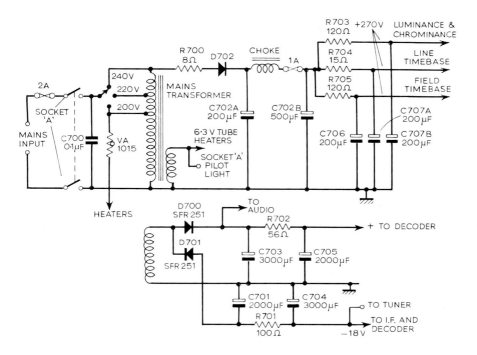

Fig. 7.21. *Power supply circuit (Decca CTV25 series).*

on the mains transformer, feeding bridge rectifiers (Fig. 7.22). The mains input primary winding is equipped with two voltage adjusters, making it possible to select a tapping corresponding closely to the input voltage at peak viewing times, thereby ensuring optimum regulator action and picture tube life. It is thus highly important for the mains tappings to be accurately selected—using an a.c. voltmeter to check the supply—at the time of installation.

The secondary supplies are 55 V, 60 V and 230 V, and the resulting stabilised and regulated d.c. outputs at the lower voltages energise the e.h.t., line timebase and field/sound modules, while the higher voltage output (270 V d.c.) feeds the video stage which, of course, demands this higher voltage to yield swings of video signal which are sufficiently high to drive the picture tube fully.

Zener diodes W9 and W10 (Fig. 7.22) stabilise the input from the regulator and produce outputs at 30 V. One of these feeds the chroma module only, allowing this to be withdrawn if necessary without affecting the monochrome performance of the set. This can be useful during a servicing operation.

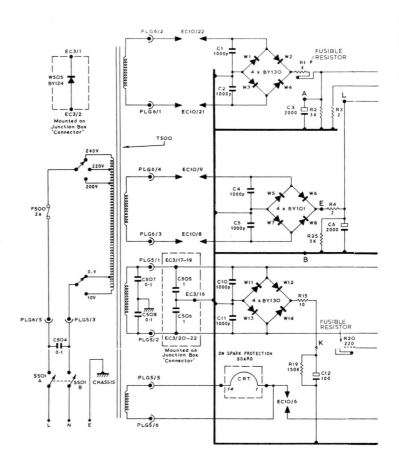

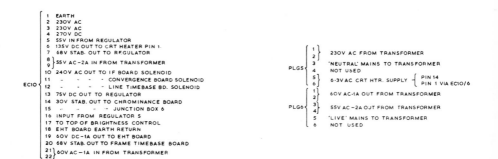

EC10
{
1 EARTH
2 230V AC
3 230V AC
4 270V DC
5 55V IN FROM REGULATOR
6 135V DC OUT TO CRT HEATER PIN 1.
7 68V STAB. OUT TO REGULATOR
8
9 } 55V AC –2A IN FROM TRANSFORMER
10 240V AC OUT TO IF BOARD SOLENOID
11 „   „   „   „ CONVERGENCE BOARD SOLENOID
12 „   „   „   „ LINE TIMEBASE BD. SOLENOID
13 75V DC OUT TO REGULATOR
14 30V STAB. OUT TO CHROMINANCE BOARD
15 „   „   „   „ JUNCTION BOX 6
16 INPUT FROM REGULATOR S
17 TO TOP OF BRIGHTNESS CONTROL
18 EHT BOARD EARTH RETURN
19 60V DC–1A OUT TO EHT BOARD
20 68V STAB. OUT TO FRAME TIMEBASE BOARD
21
22 } 60V AC –1A IN FROM TRANSFORMER
}

EC3   1,2,16—22   JUNCTION BOX TERMINATIONS

PLG5
{
1
2 } 230V AC FROM TRANSFORMER
3 'NEUTRAL' MAINS TO TRANSFORMER
4 NOT USED
5
6 } 6·3V AC CRT HTR. SUPPLY { PIN 14 / PIN 1 VIA EC10/6
}

PLG6
{
1
2 } 60V AC–1A OUT FROM TRANSFORMER
3
4 } 55V AC –2A OUT FROM TRANSFORMER
5 'LIVE' MAINS TO TRANSFORMER
6 NOT USED
}

118

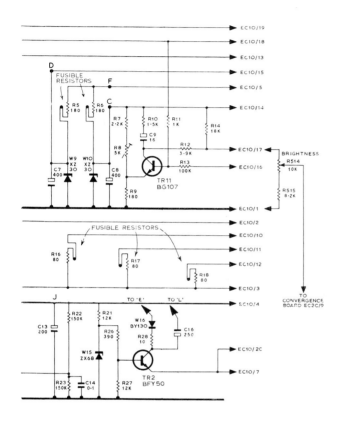

FIG. 7.22. *Power supply circuit (BRC 2000 series)*.

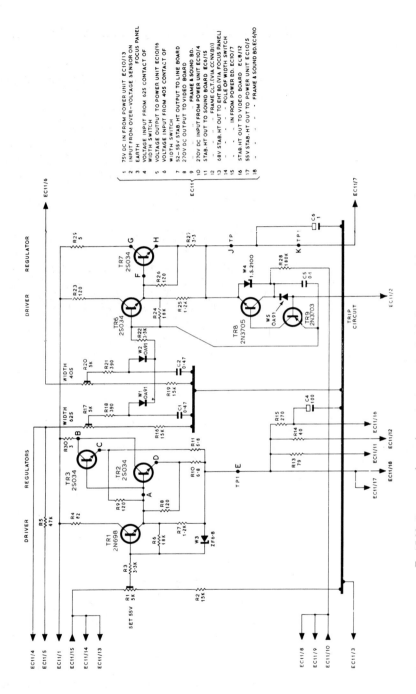

1   75V DC IN FROM POWER UNIT EC10/13
2   INPUT FROM OVER-VOLTAGE SENSOR ON
3   EARTH   FOCUS PANEL
4   VOLTAGE INPUT FROM 625 CONTACT OF
    WIDTH SWITCH
5   VOLTAGE OUTPUT TO POWER UNIT EC10/16
6   VOLTAGE INPUT FROM 405 CONTACT OF
    WIDTH SWITCH
7   52-55V STAB. HT OUTPUT TO LINE BOARD
8   270V DC OUTPUT TO VIDEO BOARD
9   ----- FRAME & SOUND BD.
10 270V DC INPUT FROM POWER UNIT EC10/4
11 STAB. HT OUT TO SOUND BOARD EC6/15
12 ----- FRAME CLT. (VIA CC1/NKB1)
13 68V STAB. HT OUT TO EHT BD.(VIA FOCUS PANEL)
14 ----- POLE OF WIDTH SWITCH
15 ----- IN FROM POWER BD. EC10/7
16 STAB. HT OUT TO VIDEO BOARD EC8/12
17 55V STAB. HT OUT TO POWER UNIT EC10/5
18 ----- FRAME & SOUND BD.EC6/10

FIG. 7.23. Power supply regulator circuit associated with the circuit of the power supply in Fig. 7.22.

Zener diode W15 produces a reference potential at the base of the stabilising transistor TR2, which delivers at its emitter a stabilised 65 V supply for the power regulator and field timebase modules.

TR1 in Fig. 7.22 is the 'sensing transistor' shown in Fig. 7.16 and referred to earlier. This safeguards the picture tube against excessive beam current, and thus avoids undue heating of the shadowmask, which could otherwise result in its physical distortion. An important component in the operation of this circuit is the 2Ω resistor (not shown in Fig. 7.22) connected between the earth and negative return line of the e.h.t. generator (see Fig. 7.16). When the tube beam current rises, the voltage across the resistor rises in sympathy, and reflects an increase in forward current into TR1 base.

This pushes up the collector current, and since the emitter/collector circuit is effectively in parallel with the brightness control, the increase in beam current is accompanied by a fall in the potential across the brightness control. This automatically increases the tube bias and thus compensates for the rise in beam current.

There are two regulator systems in Fig. 7.23, one comprising the driver TR1 and the two regulators TR2 and TR3 for feeding sections of the field timebase, sound channel and video circuits, with an input to the power supply, and the second comprising the driver TR6 and regulator TR7 for feeding the line timebase.

The output voltage of the first is adjusted by the 'set 55 V' preset, which regulates the stabilised input voltage (from the power supply) fed to the base of TR1. The use of two regulator transistors here is an artifice for producing a very low effective supply impedance.

The output voltage of the second is regulated by the 405-line and 625-line width presets. These presets, in fact, are those which equalise the line amplitude over the two standards, and since e.h.t. generation is separate from the line scan this sort of width control is quite feasible; it is also efficient since it avoids energy losses in width inductors etc.

FAULT PROTECTION

TR8 and TR9 together form a 'trip circuit', which automatically 'short circuits' the regulator transistors, TR6 and TR7, in the event of an overload. This can happen when the input current to the timebase transistors rises to a dangerous level due to a fault. Under this condition the voltage across TR6, TR7 and R27 rises above a predetermined threshold (about 10 V), and is communicated to TR8 and TR9 as a trip potential. These two transistors then conduct heavily and bypass the current from the regulator, thereby safeguarding the timebase transistors. To be successful this protective device has to come on very quickly, and the operating time of the circuit described is less than 1 ms.

The cut-outs need to be reset by pressing and releasing a single trip-lever. A fault condition is signified if a reset cut-out fails to restore the operation of the circuits.

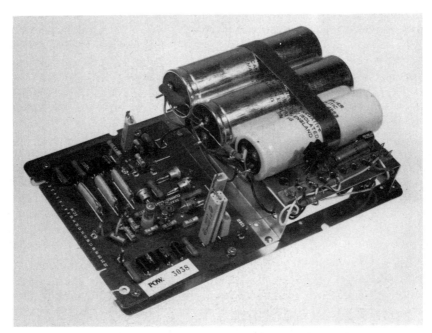

FIG. 7.24. *Power supply module. Note the vertically mounted fusible resistors (courtesy, British Radio Corporation Ltd.).*

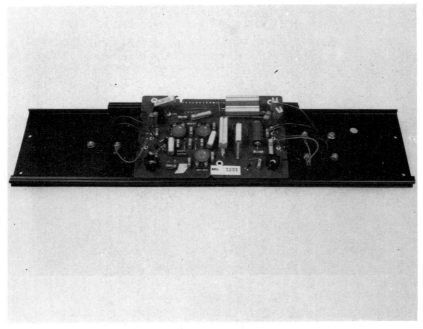

FIG. 7.25. *Power supply regulator module. The 'set 55V' and width presets are visible on the left of the board (courtesy, British Radio Corporation Ltd.).*

The mains input of the power supply is protected by a 2 A, delay-type fuse, while secondary circuit protection is given by fusible resistors, indicated on the circuit in Fig. 7.22. These need to be resoldered to restore circuit continuity after first clearing the fault condition responsible for the overloading fusing action. The power supply and regulator modules described are shown in Figs. 7.24 and 7.25 respectively.

SINGLE-STANDARD DEVELOPMENTS

Single-standard sets are now employing e.h.t. trebler units or modules, and such a device in the Decca range of single-standard models is shown in Fig. 7.26. This picture also shows the line output transformer, the line output valve and the efficiency diode, the whole lot normally being screened. The e.h.t. trebler takes the place of the usual monochrome e.h.t. rectifier valve, taking one pulse input from the line output transformer and delivering up to 25 kV of e.h.t. voltage for the picture tube final anode. Sometimes the trebler also delivers

FIG. 7.26. *E.h.t. trebler module in the Decca single-standard receivers.*

a suitable potential for the focus electrodes. The device is to be found in fully transistorised models, using transistors for the line output stage (as in the BRC 3000 series), as well as in hybrid models, like the Decca illustrated, which uses valves in the line output stage.

The trebler module contains five solid state e.h.t. rectifiers and four capacitors, the fifth, required for this type of circuit, consists of the capacitance of the picture tube between the inner and outer conductive coatings, with the glass as the dielectric, forming the e.h.t. reservoir capacitor. The basic circuit is shown in Fig. 7.27(a) and a practical circuit based on the same principle in Fig. 7.27(b). It is best to consider the operation from the basic circuit.

During positive-going line retraces pulses D1 directs peak charging current to C1 and D2 communicates the charge to C2 during the line scan periods.

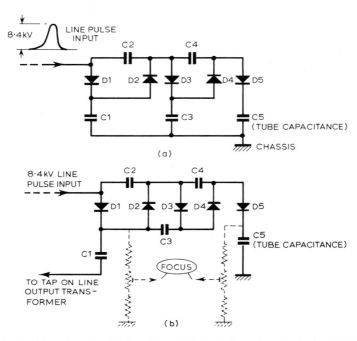

FIG. 7.27. *Basic e.h.t. trebler circuit (a) and practical circuit (b). The action is described in the text.*

Also during retraces D3 directs peak charging current to C3, but since this is via C2, C3 charges to twice the peak value through D4. Similarly, during retraces D5 charges the tube capacitance to the peak value of the pulse *plus* two-times the voltage held by C4, so the tube capacitance charges to three times the peak voltage of the retraces pulses, which is to about 25 kV.

The peak voltage across all the capacitors in the trebler proper with the exception of C3 is a third of the total e.h.t. voltage. The peak voltage across C3 is 2V/3, where V is the total e.h.t. voltage, but this can be arranged to match the peak voltage across the other components, allowing them all to have the same voltage characteristics, by altering the circuit to that in Fig. 7.26(b). One difference in this practical circuit is that the bottom plate of C1 is returned to a 'tap' on the line output transformer instead of to chassis, and by selecting a suitable 'tap' the e.h.t. voltage can be adjusted to the required value without difficulty. For example, if the 'tap' carries a negative-going pulse, the e.h.t. voltage will be increased by twice the peak value of the pulse. This happens because C3 is connected to the junction of D1 cathode and D2 anode, rather than to the chassis, which represents another difference from the basic circuit.

In principle, the peak current through C1 and C2 is twice that through the other arms of the circuit, which means that in theory at least both C1 and C2 should be twice the value of C3 and C4 to get the same volts drop across them; but in practice the circuit works adequately when all the trebler capacitors are of like value, this being of the order of 1 nF to 1·5 nF.

Focus potential can be picked up from the first stage of the trebler or from a bleeder across the total e.h.t. supply, as shown by the dotted circuits in Fig. 7.27(b). It is noteworthy that the resistors need to be of significant value to drop the voltage sufficiently, which means that they are vulnerable to value change.

For the best e.h.t. regulation the retrace pulse must have the longest possible duration, and the line output stage driving a trebler is generally designed to cater for this condition, one scheme being to tune the transformer to the fifth harmonic of the retrace frequency. In practice a regulation better than 2 megohms can generally be attained.

E.H.T. trebler modules cannot easily be repaired in the workshop. The best plan is to keep a replacement or two in hand for the models handled and return the faulty ones to the manufacturer.

'CHOPPER' CIRCUIT STABILISATION

It has already been noted that fully transistorised sets require good supply regulation, and details of the 'regulator' in the BRC 2000 series have been examined. A different arrangement is used in the single-standard BRC 3000 series. Three separate d.c. supplies are obtained from an autowound mains transformer and rectifiers, the transformer carrying also a small winding of 6·3 V r.m.s. for the picture tube heaters.

The stabilised high current supply required by the line, field and sound output stages is yielded by a 'chopper' supply. The main element in this circuit is a high power transistor arranged to feed pulses at line frequency into an inductive reservoir which in turn feeds the output load. The circuit operates by virtue of the *duration* of the pulses being caused to increase or decrease as the load varies. Half-wave rectified d.c. is fed to the collector of the 'chopper' transistor which is switched on and off at line frequency, and a feedback circuit, sensing the change in load conditions, continuously varies the mark–space ratio of the pulses so that the power fed into the reservoir matches exactly the requirements of the circuits under changing conditions of load. The feedback circuit, while stabilising the output voltage, also smoothes the 50 Hz ripple. Operational source impedance at the output is less than 1Ω.

The trigger pulses for the 'chopper' circuit are obtained from the line time-base generator, itself powered from a 30 V stabilised line, provided by a series transistor stabilising stage. Zener diodes are also employed in the circuit for supply control and regulation. The trigger pulses operate a monostable circuit whose action is delayed until the 30 V stabilised supply is established. This en-sures that the circuits being fed from the supply are not over-run immediately following switch-on. Dynamic and excess voltage trips are included in the overall circuit for safety.

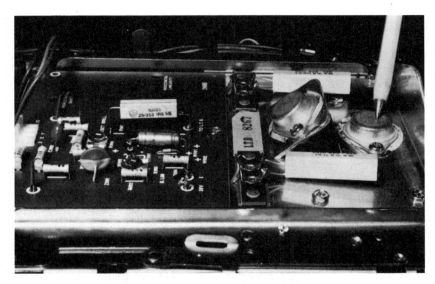

Fig. 7.28. *Heat-sink-mounted line output transistors in the BRC single-standard 3000 series. The line timebase board is on the left-hand side of the power transistors.*

The line timebase in the 3000 series, as in the 2000 series, is transistorised, and Fig. 7.28 shows the pair of heat-sink-mounted power transistors to the right-hand side of the timebase board.

CHAPTER 8

Luminance and Colour-Difference Amplifiers and Grey-Scale Tracking

WE NOW HAVE an appreciation of how the shadowmask tube is able to produce three rasters in red, green and blue separately or mixed and how these can be adjusted to yield a display, not only of almost any hue in nature, but also from black through the whole range of greys to peak white. We also have some idea of the requirements for purity and convergence and how these are satisfied in practice. This chapter first investigates the circuits connected to the base electrodes of the shadowmask tube, looking particularly at the biasing of the three guns and at the colour-difference and luminance signals applied to them, and finally deals with the very important consideration of grey-scale tracking.

We have seen that the colour camera delivers signals corresponding to the three primary colours along with the luminance or Y signal. Processes of encoding translate the primary colour signals first into colour-difference signals and then into chroma signal, and the chroma signal is that which is eventually delivered to the receiver's decoder. From this the decoder yields the original colour-difference signals which, after amplification and $G - Y$ matrixing, are applied to the tube guns along with the detected Y signal.

DERIVING THE PRIMARY COLOUR SIGNALS

The tube has to be operated by the three primary colour signals corresponding to those originally produced by the camera. There are two ways of achieving this. One takes the colour-difference signals and feeds them to the red, green and blue grids (separately) while a $-Y$ signal is applied to the three cathodes together.

Each original primary colour signal now modulates its own beam because the phases of the colour-difference and luminance signals are such that the $-Y$ component in each colour-difference signal is cancelled. The idea is illustrated in Fig. 8.1. Because a signal on a cathode has the opposite effect (on beam current) of a similar signal on the grid, the $-Y$ signal on the cathode cancels the $-Y$ component of each colour-difference signal, thereby leaving only the primary colour signal modulating the beam. For example, R results from $R - Y$ because $(R - Y) + Y$ equals R, or because $(R - Y) - (-Y)$ equals R.

The alternative scheme arranges for the primary colour signals to be produced from the colour-difference signals and Y signal in a matrix prior to the guns of the shadowmask tube.

DERIVING G − Y

In both of these schemes, the green colour-difference signal, which is not transmitted, is obtained in a matrix (see Fig. 8.2) following the PAL decoder.

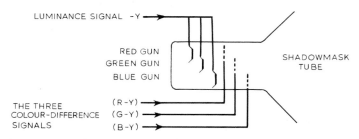

FIG. 8.1. *Showing how each gun 'sees' only its corresponding primary colour signal. This is because the Y signal at the cathodes cancels out the Y part of the signal at each grid, leaving only the primary R, G and B signals. For example, (B − Y) − (− Y) = B. The same applies to the other two colours. In some sets this 'matrixing' takes place in transistors, and then the tube cathodes receive only the primary colour signals.*

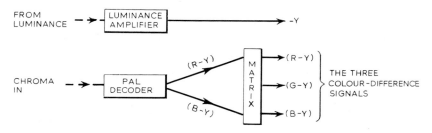

FIG. 8.2. *The red and blue colour-difference signals are matrixed to produce the missing green colour-difference signal at the set. G − Y is equal to −0·51(R − Y) −0·19(B − Y), as explained in the text.*

This is possible, as we have seen, because all three chroma signals are already represented in the Y (luminance) signal.

Let us recall that $Y = 0·3R + 0·59G + 0·11B$. What we wish to get from the matrix is G − Y, since we already have R − Y and B − Y. Rounding off the figures, and doing a bit of mathematical manipulation, we can start by subtracting Y from both sides of the fundamental expression thus:

$$0 = 0·3(R − Y) + 0·6(G − Y) + 0·1(B − Y)$$

Then, and noticing the change of 'signs',

$$0·6(G − Y) = −0·3(R − Y) − 0·1(B − Y)$$

Finally, dividing out the 0·6 gives the approximation

$$G - Y = -\tfrac{1}{2}(R - Y) - \tfrac{1}{6}(B - Y)$$

What the matrix amounts to is nothing more than a resistive network arranged to take red and blue colour-difference signals in the approximate ratios of $\tfrac{1}{2}$ and $\tfrac{1}{6}$, change their phase (as received from the decoder) and add them.

Another way is as follows:

$$R - Y = 1·0R - (0·3R + 0·59G + 0·11B)$$

$$= 1·0R - 0·3R - 0·59G - 0·11B$$

$$= 0·7R - 0·59G - 0·11B$$

$$B - Y = 0·3R - 0·59G + 0·89B$$

$$G - Y = 0·3R + 0·41G - 0·11B$$

Thus in terms of colour-difference factors we get

$$G - Y = -0·51(R - Y) - 0·19(B - Y)$$

Matrixing generally takes place somewhere in the colour-difference pre-amplifier stages, and several arrangements are in current use. The block diagram in Fig. 8.3 gives an impression of the general set-up and indicates possible points for matrixing $R - Y$ and $B - Y$ to yield $G - Y$.

D.C. CLAMPING

Unlike the video amplifier in most monochrome sets, colour-difference amplifiers use a.c. coupling. This makes d.c. clamping or restoration necessary at each grid as in Fig. 8.3. Since there are three guns to contend with, it is essential to zero-clamp each of the three colour-difference signals to ensure a stable grey-scale setting of the tube biasing.

A popular arrangement in valved sets takes the form of a triode pentode in each colour-difference amplifier circuit, the pentode being the colour-difference

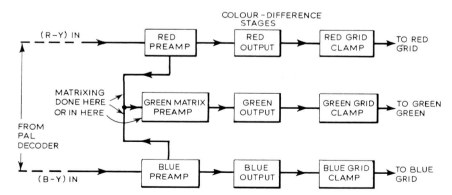

FIG. 8.3. *Block diagram of red, green and blue colour-difference amplifiers. G − Y phasing depends on colour-difference circuits.*

output valve and the triode the clamp. This sort of clamp is a little different from the simple d.c. restoration diode in that it is 'triggered' by pulses from the line timebase. These pulses are fed to the clamp triode grid. As they are about 50 V positive-going they switch the valve hard-on and cause it to saturate or 'bottom'.

A colour-difference channel (the red one) rear-end is shown in Fig. 8.4. The green and blue ones are identical. When a line pulse occurs the triode anode, and hence the tube grid, is clamped to the potential at the triode cathode, and

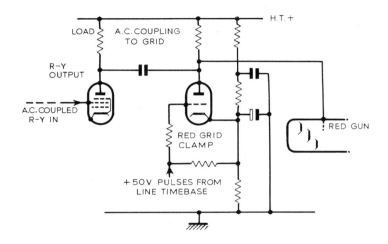

FIG. 8.4. *Red colour-difference output stage and triode clamp. The green and blue stages are similar.*

this is arranged to correspond to zero composite signal video level. This level is held during the effective discharge time of the coupling capacitors which, in general, is about 6 ms, relatively long compared with the time of one line scan.

It will be recalled that the chroma signal is limited in bandwidth. This means that the colour-difference amplifiers driving the tube grids—and, indeed, the chroma amplifier feeding the chroma detectors—need pass signal components extending up to only 1 MHz. The chroma amplifier is generally double-side-band tuned, having a passband of ± 1 MHz between 3 dB points.

LUMINANCE AMPLIFIER

The luminance amplifier is virtually the same as the video amplifier in monochrome sets. It has a wider bandwidth than the colour-difference amplifiers and usually employs a separate valve (rather than the pentode of the triode-pentode valve). Like the red, green and blue colour-difference amplifiers, however, it may be fed from a preamplifier section using a mixture of valves and transistors. This is particularly so with hybrid sets and dual-standard sets which need video phase-changing on switching the standard.

The luminance output stage in common use is shown in Fig. 8.5. The valve is a high-slope pentode designed specially for video applications and its anode load is R1 (about 2·7 kΩ). This gives a bandwidth of the order of 6 MHz at −3 dB points. The ordinary peaking coil is L1 (keeping up the top response) while R2 and C1 are decoupling components.

LINE AND FIELD BLANKING

The cathode circuit is interesting since it contains a transistor. This is virtually a switching device for line and field blanking (in monochrome sets such pulses come in on the picture tube grid). The transistor is held conducting (bottomed) by the positive potential at its base through R3. When a line or field pulse occurs, the transistor is switched off, the valve is 'open-circuited', the anode

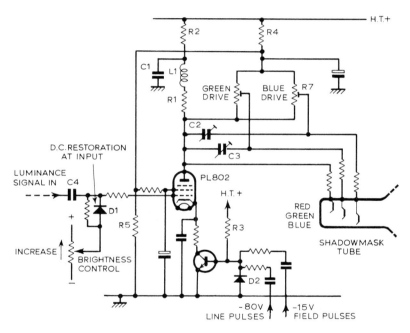

FIG. 8.5. *Luminance output stage, showing video drive presets and line and field blanking by the transistor in the cathode circuit.*

voltage rises and the tube guns are cut-off. The signal at the valve anode, therefore, contains positive-going line and field blanking pulses, which are reflected as negative-going at the tube grids.

Before we consider the circuits to the tube control grids, let us investigate the anode feeds to the tube cathodes. It is not possible simply to parallel the three cathodes to a common point in the video valve anode circuit. This may seem all right in theory, but in practice it is necessary to take into account the unequal light efficiencies of the three colour phosphors on the screen. The video drive to each cathode is thus separately adjustable. This also facilitates grey-scale tracking adjustments, considered later.

VIDEO DRIVE PRESETS

The complicated anode circuit in Fig. 8.5 provides preset adjustments on the green and blue cathodes. A fixed black-level reference voltage is established by the potential divider (R4, R5) across the h.t. supply while the green and blue drives are under the control of potentiometers R6 and R7 respectively. The red cathode is connected to the anode direct.

At full-drive (black-level), therefore, the voltages each side of the preset are equal. At drives from black towards peak white the anode potential falls and so the required drive potential can be picked up from the sliders. The potentiometers allow the actual video drive to the tube cathodes to be adjusted for correct grey-scale balance over the range of brightness control and contrast (referred to again later).

H.F. COMPENSATION

High-video-frequency compensation is provided by trimmers C2 and C3 respectively for the blue and green cathodes. These simply shunt the potentiometers. Compensation for d.c. component loss in the luminance amplifier takes place in the screen-grid circuit which introduces a degree of d.c. degeneration.

BRIGHTNESS CONTROL AND D.C. RESTORATION

Monochrome-set brightness is adjusted by a potentiometer in the tube grid circuit, this just giving an adjustment of grid voltage. This cannot be used in colour sets so easily because the grids are connected direct to the clamp-triode (Fig. 8.4). Another method is adopted and this works in conjunction with the luminance amplifier (Fig. 8.5).

The luminance signal is a.c.-coupled to the control grid through C4 and the d.c. component is restored by diode D1, which operates on the tips of the sync pulses. The d.c. on the control grid is adjustable by the brightness control, as the circuit reveals. When the negative voltage is increased, anode current falls and anode voltage rises. This causes all the tube cathodes to go more positive, reducing the current in all three beams together and turning down the overall brightness.

On a monochrome picture the red, green and blue colour signals fall to zero. Under this condition the three grids are clamped to zero colour-difference signal level by the clamping triodes. One can now better appreciate the need for accurate grid clamping. A drift in grid potential would soon put the three beams out of track and cause the compounded raster (or picture) to change sadly in hue on a monochrome transmission, in particular, and also when the brightness control is operated. Trouble in the triode-pentode valves can cause this sort of symptom (e.g. grid transmission) and inter-electrode leakage.

On a monochrome transmission, the grids are clamped to a fixed potential and the signal swings are applied only to the three cathodes. A swing of about 100 V between black level and peak white is necessary for maximum contrast. This is relative to the fixed black-level potential which, in turn, is related to the zero colour-difference potential on the grids themselves.

With the Mullard PL802 luminance amplifier valve correctly set up, full-drive or black level (negative picture modulation on the 625-line standard) corresponds to about 220 V on the anode. The value of negative voltage required on the grids of the guns for beam current cut-off depends on the first-anode potentials (hence the need to adjust these for correct grey-scale tracking). However, a cut-off potential of about −110 V at the grids is initially preferred. Since the cathodes at black-level are +220 V, the grid clamps must be returned to +110 V. Each gun then 'sees' −110 V on its grid relative to cathode (beam cut-off having been set by the first-anode presets).

However, for correct grey-scale tracking, the blue and green cathodes have respectively about 94 and 96 per cent of the potential of the red cathode. Spreads in tubes, of course, can alter these ratios. The point is that the amount of beam current needs to be proportioned relative to each gun so that white is produced on the screen. White is not produced when the three beams carry the same current.

The luminance amplifier in Fig. 8.5 has an overall gain of about 35 times, which means that a 3 V black-to-white input will give approximately 100 V black-to-white output, representing full luminance (or chrominance) drive. The red, green and blue colour-difference amplifiers have to provide somewhat higher signal swings than the luminance amplifier. This is due in part to the lower sensitivity of grid drive compared with cathode drive. Allowances also have to be made for tube and component spreads and also for the fall in tube sensitivity as it ages and the PAL signal weighting factors (see Chapter 10 and below).

Maximum grid drives are thus in the order of 183 V, 102 V and 218 V (all peak-to-peak) respectively for red, green and blue. The B − Y channel gain is particularly high because the transmitted blue chroma signal is purposely reduced to 56 per cent of the red chroma signal, and the corresponding amplifier has to provide the necessary gain compensation. Amplifier gains run up to 800 times in the colour-difference channels and adjustment is provided in the preamplifier stages for balancing.

133

Because the red, green and blue lights on the shadowmask screen are governed by the currents of the corresponding electron beams, and because white light is discerned by the eye only when the energies of all three colours are about equal, it follows that the beam currents must change from zero to maximum in perfect unison when the brightness control is operated to prevent colour contaminating a monochrome display. The same applies, too, when the three beams are under Y-signal control (monochrome transmission) with the three grids clamped to black-level.

The tracking of the three beams over the full range of luminance has something in common with the tracking of three tuned circuits in a superhet receiver (r.f., mixer and local oscillator) over the whole tuning range. In a superhet receiver we have—in principle at least—two adjustments for each tuned circuit, a trimmer for the h.f. end of the band and a padder for the l.f. end. Likewise, each beam of the shadowmask tube has two adjustments, one for the low-brightness end of the range and one for the high-brightness end.

The high-brightness adjustments, which are the video drive presets, are shown in the complete luminance output stage circuit in Fig. 8.5. Actually, there are two, not three, presets in that circuit. The red cathode is directly connected to the luminance amplifier anode. This gives one fixed point, allowing the green and blue cathodes to be adjusted relative to this. Some sets, however, do have also a preset or tapped potential-divider for the red cathode.

We need to keep in mind that similar electron beam currents do not give red, green and blue lights of equal energies. This is because the lumen efficiency differs between phosphors of different colours. Red is the least efficient, and so the red beam needs greater current than the two other beams for the same light energy at the phosphors. For this reason the red cathode is generally connected direct to the luminance amplifier output.

It will also be recalled that although white light is produced when the red, green and blue light energies are equal, the eye does not see these equal-energy lights as lights of equal brightness. Yellow-green seems the brightest, with red and blue falling at the sides of the luminosity curve.

Having set-up the three beams and their tracking, the correct balance could then be disturbed by changes in potential on other tube electrodes. The first and final anode potentials are particularly critical in this respect, and this demands good supply regulation, as was explained in Chapter 7.

Grey-scale tracking, then, is obtained at the high brightness end of the scale by the video drive presets to the tube cathodes. The low brightness end is related to the cut-off characteristics of the electron guns and we need to adjust something here that affects this characteristic—preferably the first anodes.

When adjusting monochrome sets having something wrong with the range of the brightness control, it is sometimes discovered that the reason for the beam current cutting-off too early when the brightness control is retarded (conversely, screen illumination being inadequate with the control fully turned up) is low first anode potential. This also affects the focus of the scan spot.

Replacing an increased value first-anode feed resistor or 'leaky' decoupling capacitor restores the first-anode potential and the normal range of brightness control. Reduced first-anode potential effectively shortens the gun grid-base.

Presetting this potential separately on each gun of the shadowmask tube makes it possible to 'match' the grid bases of the three guns at the low beam current end of the scale to attain the best balance at this point. When these presets are adjusted, and also the cathode drive presets, it is possible to get the tracking to hold *fairly* accurately at all intermediate levels of brightness. The characteristic of the three guns then 'match' reasonably well.

It should also be borne in mind that tracking maladjustment can affect a colour picture, causing the hues to change when the brightness of all or part of the scene changes.

An overall view of a grey-scale tracking circuit is given in Fig. 8.6. The feeds to the three first anodes are shown, but for the sake of simplicity only the blue

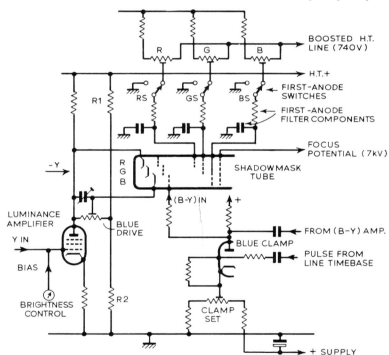

FIG. 8.6. *Skeleton tube control circuit, showing the adjustments for grey-scale tracking. Note that a similar 'drive preset' exists for the green cathode. The circuit also shows the B − Y input to the blue grid, and the blue clamp triode.*

drive preset and clamp triode are included. We have already seen that the triode clamps and drive presets are virtually identical on each colour.

R1, R2 potential divider across the h.t. supply establishes the black-level reference. This simply means that when the luminance amplifier is biased for black-level, the voltage at R1, R2 junction is the same as that at the luminance valve anode.

135

Grey-scale tracking adjustments

Amplifier biasing is affected by the brightness control, as this varies the control grid voltage. Thus, we must establish the correct setting for this control by adjusting until the luminance anode voltage equals that at R1, R2 junction. This is the black-level setting. Now, at this setting we need a test signal carrying shades of grey (i.e. a staircase waveform). We must carefully check the displays of grey, with the set switched to monochrome, and adjust the drive presets to remove colour from the lighter greys and the first anode presets to remove colour from the darker greys. To clear colour from over the whole range of contrast, from black to white, some compromise adjustment may be needed, and absolute perfection is rarely possible.

Some sets have a 'clamp-set' preset which sets the clamp potential at the grids. When the correct black-level setting of the brightness control has been found, as just explained, screen illumination (all colours equally) should just be extinguished. Since the clamp potential determines the gun biasing, it is obvious that the clamp-set affects the cut-off. The idea, then, is to adjust this preset for the cut-off point at the black-level setting of the brightness control.

In some setting up procedures, the clamp-set preset is brought into play in conjunction with the green and blue drive presets on the lighter grey sections of the contrast pattern.

When the dynamic grey-scale tracking is correct (i.e. on colour bars from the transmitter or pattern from service generator with the set's colour turned down or switched off), the tracking will be correct over the range of manual brightness control. This is obvious, of course, since the source of control is at the luminance valve control grid on both dynamic and manual functions.

Incorrect manual tracking is observed when the unmodulated monochrome raster (set turned to monochrome) changes, say, from yellow through white to blue as the brightness control is advanced. This is very disturbing! It is best to undertake the tracking adjustments in a darkened room to ensure that colour contamination does not creep in at extremely low levels of brightness, which would fail to be discerned in a room of normal illumination.

Different sets have slightly different arrangements for grey-scale tracking. One method which is somewhat different, and which is used for adjusting the grey-scale of the dual-standard BRC 2000 series and the single-standard 3000 series, involves collapsing the field and then adjusting the low-level brightness of the red, green and blue lines. All sets have three first-anode presets but some have four drive presets, two for blue and one each for red and green. The Philips G6 chassis, for instance, has presets like this, and a special procedure is laid down for their adjustment. These drive presets adjust the 'white tones'.

The 'grey tones' are adjusted by the three first-anode presets in conjunction with a 'luminance switch'. This disconnects the Y signal at the cathodes. In that condition, each first-anode preset is adjusted in turn on the red, green and blue rasters in a darkened room for almost complete cut-off (i.e. until the raster is barely perceptible). A particular colour raster is obtained by switching off the other two guns by the first-anode switches, shown in Fig. 8.6.

Finally, the three guns are switched on together, and also the luminance switch. The adjustments are then checked on a step-pattern or test card which has a good neutral grey tone. If a touch-up adjustment is required to optimise the greys, it is permissible to adjust any two of the first-anode presets slightly.

The clamp-set (called 'c.r.t. cut-off adjustment' on this model) must be adjusted first. A high-resistance voltmeter is used to measure the voltage on the blue cathode with respect to chassis and next on the blue grid. The latter is adjusted, by the clamp preset, until it is 110 V less than the voltage on the cathode. This ensures that the blue gun, at least, is biased to 110 V negative on its grid (for cut-off).

Some sets have a colour on/off switch or button which kills the chroma for monochrome adjustments of this nature, while on some other models the colour is turned down merely by fully retarding the colour control.

At this juncture it will be instructive to run through the procedure for grey-scale tracking the BRC 2000 series sets, in conjunction with the video module circuit diagram in Fig. 8.14, and the module picture in Fig. 8.15. Note that the final setting-up procedure should be undertaken on a 625-line channel, preferably on a monochrome transmission. Initial adjustments are made on horizontal lines with the field collapsed, leaving horizontal lines.

1. Set the tint control to range centre and turn the three first anode presets on the convergence board to minimum. These are R33, R35 and R36 in Fig. 6.26.
2. Connect a 20,000 Ω/V testmeter set to the 250 V full-scale range to Test Point 2 (T.P.2) on the video module circuit with the positive lead of the meter to T.P.2 and the negative lead in turn across the collector loads of the red, green and blue video output transistors.
3. Adjust each of the bias presets (R42, R61 and R79) for minimum d.c. reading on the testmeter.
4. Set the three drive presets (R38, R58 and R75) to their mid-positions.
5. Operate the 'set white' switch (S1A/B1C) on the convergence board (Figs. 6.26 and 6.27) to collapse the field. (Note: This is a three-section switch which cuts off the power supply to the field timebase and presets the brightness of the horizontal lines to a low level to avoid screen burns.)
6. Transfer the positive of the testmeter to Test Point 1 (T.P.1) on the video module circuit (i.e. the base circuit of the luminance emitter-follower, TR3) and the negative lead to chassis.
7. Adjust the video reference preset (R14) on the convergence board for a reading of 9·5 V on the testmeter.
8. Re-connect the testmeter as for (2) and adjust in turn each bias preset for 90 V.
9. Connect the meter positive lead to tag 12 (all the grids in parallel) on the tube base connector and the negative lead to the chassis. Adjust the c.r.t. grid bias preset (R30 in Fig. 7.10) on the field timebase and sound module for 30 V.
10. Slowly turn up each first-anode preset in turn until the three colours are just visible as three equally bright horizontal lines on the screen. The three

first-anode beam switches (S2, S3 and S4) on the convergence board may be switched on and off in turn on the individual beams for comparison purposes. However, if any of the horizontal lines *fail to appear* with the corresponding first-anode preset at maximum, the following procedure should be adopted:

    (*a*) Leave the first-anode preset(s) corresponding to the absent colour(s) at maximum.

    (*b*) Set the first-anode potentiometer(s) controlling the *visible* line(s) to minimum.

    (*c*) Advance the c.r.t. grid bias preset (R30) on the field timebase module until the previously absent colour(s) are just visible.

    (*d*) Set all three first-anode presets to a minimum and repeat (10).

11. With all three beams switched on, restore the field with the 'set white' switch on the convergence board.

12. Adjust contrast and brightness controls for a good monochrome, 625-standard picture or Test Card.

13. If necessary, re-adjust the video gain presets to improve the highlights and trim the first-anode presets to remove any tinting in the shadows.

TINT CONTROLS

A preset labelled 'tint' is found in some models and the function of this in the Philips G6 chassis is shown in Fig. 8.7. It is located in the blue colour-difference output stage cathode but also connected to the red colour-difference output

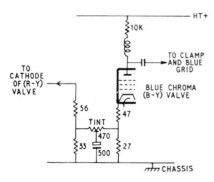

FIG. 8.7. *'Tint control' (Philips G6 series). This is explained in the text.*

valve cathode. The slider of the control applies cathode compensation (by the 500 µF electrolytic to chassis) differentially between the red and blue amplifiers.

On adjusting the slider towards the blue amplifier cathode, the gain of the blue amplifier is increased and the red amplifier decreased, and vice versa. This is to balance the red and blue gains to give the most desirable flesh tones in a colour picture without, of course, affecting the luminance (black-and-white).

The 'personal tint' control fitted to some sets is different again. This allows the viewer to adjust the quality of the white light given by the tube when the set is receiving monochrome. The adjustment goes from bluish-white, through the standard illuminant D to sepia-white. The control can also serve to neutralise tracking errors between the three electron guns which may develop after a period of use, and thus possibly delay a service call for the full grey-scale tracking treatment. The control works by providing a differential adjustment to the cathodes of the shadowmask tube, rather than to the grids, as with the tint control previously explained.

We have seen that to control the brightness the standing potential at the luminance amplifier valve anode has to be adjustable. This, as already shown, can be secured through the brightness control varying the valve grid bias or the transistor base bias. Another way is by varying the valve screen-grid potential and this is adopted in the Philips chassis, as shown in Fig. 8.8. Changing the screen-grid potential alters the valve conductivity, so to speak, and swings the voltage at the anode.

Figure 8.8 also has a bias preset. This comes in only on 625 lines. The idea is first to adjust for a normal 405-line picture, switch to a 625-line transmission and then to adjust the bias preset for the same brightness of picture. The zener-

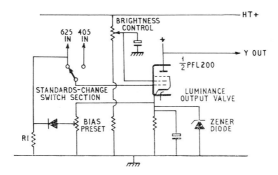

FIG. 8.8. *Control of brightness achieved by regulating the screen supply to the luminance amplifier.*

diode across the cathode stabilises the voltage here, while the ordinary diode on the bias preset slider ensures a current flow through load R1 in one direction only.

COLOUR TUBE ELECTRODE POTENTIALS

Since we are concerned at the moment with the picture tube end of the set, it is appropriate to look at the sort of potentials required by the first-anodes and focusing electrodes. The first-anodes require between 220 V and 520 V, adjustable by the grey-scale tracking presets. Since this is relative to the cathode, and because the cathode at black-level is at 220 V, the real supply range is from 440 V to 740 V. This is picked up from the boosted h.t. supply, as in monochrome sets. The supply is filtered by series resistors and shunt capacitors, and these components are shown on the first-anodes in Fig. 8.6.

Up to 7 kV is needed for electrostatic focusing. This high voltage is obtained either by rectification of the pulses at the booster diode cathode or by tapping down the e.h.t. supply. Another method uses one stage of the e.h.t. trebler. These potentials were considered in Chapter 7.

We now come to the luminance and colour-difference preamplifier stages. While valves might be used for the luminance and colour-difference output stages (since a high level of drive is needed from these), transistors are often employed for the preamplifiers feeding them.

Let us first look at the luminance preamplifier (Fig. 8.9a). The tube cathodes require a total dynamic drive (picture plus sync signal) of about 140 V for full modulation. To provide this the luminance amplifier stage requires an input of about 3 V black-to-white swing. This is obtainable from the detector or phase splitter. The preamplifier stages are concerned mainly with phasing the signals, and impedance matching.

TR1 in Fig. 8.9(a) receives the composite detector signal at its base and delivers it at both emitter and collector. The standards change switch selects the appropriate 'negative' or 'positive' output, feeding the signal to TR2 base. In most dual-standard monochrome sets the change in video modulation is accomplished by polarity reversal switching at the vision detector diode; this is avoided in colour sets, when a stage such as TR1 (Fig. 8.9a) is adopted, because of the phase reversal between emitter and collector.

The main detector can be connected direct to TR1 base. On the 405-line standard the signal phase on the emitter is the same as on the base. On the 625-line standard, however, the signal is taken from the collector, and is provided with the required phase inversion because of the negative-going video signal.

The base of TR2, therefore, receives correctly phased signals on both standards, via the standard change switch section. The video signal proper is taken from the emitter of TR2, via the contrast control, and fed to the luminance output valve, while the signal for application to the sync separator (requiring reversed phase for correct operation) is taken from the collector.

HIGH-LEVEL CONTRAST CONTROL

Maximum video signal from the contrast control is of the order of 5 V. Although the contrast control is at this relatively high signal level, the signal to the sync separator (from the collector) is completely unaffected by the setting of the contrast control.

L1, C1 circuit in TR1 emitter circuit is tuned to the chroma subcarrier (4·4 MHz) and this puts a dip in the luminance channel at that frequency, preventing the subcarrier interfering unduly with the luminance signal.

VISION A.G.C.

The d.c. voltage at TR1 emitter increases positively as the strength of the detector signal increases, and this voltage is used again for a.g.c. on both standards. A positive-going a.g.c. voltage is required because the controlled stages are transistors in hybrid sets. Control is usually applied to the first i.f. stage and the tuner r.f. stage, with a delay system ensuring that r.f. control is not applied until the i.f. stage is fully controlled. This technique provides a good signal-to-noise performance—particularly important in colour sets.

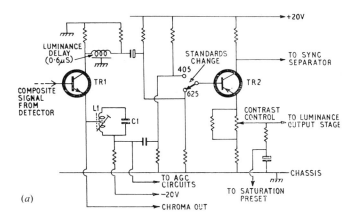

FIG. 8.9. (a) Luminance preamplifier, showing the various take-off points. (b) Typical luminance delay line.

A 625-line a.g.c. voltage is often derived from a sync pulse tip-amplitude detector, giving black-level reference. A Mullard circuit after this style has been designed to work in this way 'on tune' and to provide mean-level a.g.c. 'off tune' —that is, while the set is being tuned.

On 405 lines a gating system is invariably featured, and black-level reference is provided by the back porches of the line sync pulses in the usual manner.

ELECTRONICALLY COUPLED CONTRAST AND SATURATION CONTROLS

Returning to the luminance channel itself, while most colour sets feature a user's saturation control, some circuits adopt a saturation preset electronically geared to the contrast control. This has merit because, after contrast adjustment, a set with a separate saturation control needs adjustment here, too. The geared contrast/saturation idea avoids this because the colour is turned down automatically as the contrast is turned down, and vice versa.

A preset is sometimes fitted to secure the correct ratio of colour-to-contrast in the first place, and this ratio remains over the range of the contrast control. In Fig. 8.9(a) this is achieved by the slider circuit of the contrast control.

LUMINANCE DELAY

A major feature of the luminance channel is the delay circuit. This must not be confused with the PAL delay line. The luminance delay serves an entirely different purpose; it delays the signal in the luminance channel so that it arrives at the shadowmask tube at exactly the same time as the colour-difference signals.

This is necessary because the reduced bandwidth of the chroma circuits has the effect of delaying the signal relative to that travelling in the wider-band luminance circuits. A delay of around 600 to 830 nanoseconds satisfies the bandwidth difference between the two channels, depending on design, and in Fig. 8.9(a) this is incorporated in the 625-line coupling from the collector of TR1. Fig. 8.9(b) shows a typical delay line.

INCORRECT COLOUR REGISTRATION CAUSES

Incorrect delay results in the colour being displaced to the right or left of the corresponding luminance components in the picture. In other words, the colour fails to register correctly with the monochrome parts of the picture.

Capacitance distributed along an inductive line represents one form of delay device employed in the luminance channel. The capacitance is formed of copper strips along the entire length of the line. A fault in these or their 'earthing' can affect the loading and impedance coupling of the line (as well as changing the delay time), and the result is multiple reflections from one end of the line to the other and back again (similar to reflections in badly terminated signal feeder). This gives 'ringing' effects on the monochrome picture relative to the colour.

It is worth noting that a change in bandwidth of the luminance or chroma channel can modify the delay, again causing mis-registration on the picture. This means that misalignment can also show up as colour displacement, as well as the more familiar symptoms. As for the chroma signal, this is extracted from the emitter of TR1 (Fig. 8.9) rather than from the detector direct. This gives the required impedance match to the input of the decoder.

The colour-difference preamplifier stages in the Decca CTV25 receiver are shown in Fig. 8.10. These drive the red, green and blue output stages and contain

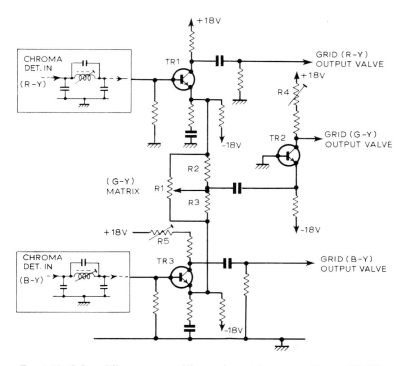

FIG. 8.10. *Colour-difference preamplifiers and matrixing system (Decca CTV25).*

the green matrixing. Signals, via low-pass filters, are fed to the bases of TR1 and TR3 from the red and blue detectors respectively, with these transistors working in the common-collector mode for matrixing and in the common-base mode for red and green drive to the output valves. That is, the drive signal is taken from the collectors and the matrixing signal from the emitters. They can thus be considered as phase splitters.

TR2 is arranged in the common-base mode, with the base 'earthed' and the input signal applied to the emitter. The input here comprises the matrixed R − Y and B − Y signals, taken from the matrix network R1, R2 and R3. The requirements for matrixing have already been described, and the resistors in the matrix are proportional, so that the G − Y signal is developed in TR2 collector circuit. The R − Y and B − Y signals are given the correct phase for matrixing by being extracted from the emitters of TR1 and TR3. Presets R4 and R5 permit adjustment to the amplitude of the G − Y and B − Y signals with respect to the R − Y signal to suit the tube drive requirements.

TRANSISTORISED BED LUMINANCE AND COLOUR-DIFFERENCE CIRCUITS

The trend towards all-transistor colour sets means that the luminance and colour-difference *output stages*, as well as the preamplifiers feeding them, must use transistors instead of valves. We have seen that the problems are easily resolved so far as the preamplifiers are concerned, but those of the output stages are somewhat more involved as they need to supply relatively high level, wide-band drive signals for the picture tube cathodes and grids.

TRANSISTORISED LUMINANCE AMPLIFIER CHANNEL

Nevertheless, contemporary transistors can provide the demands in suitable circuits. An illustration of an all-transistor luminance channel by Mullard is given in Fig. 8.11.

This is for 625 lines only, and the preamplifier stages have much in common with those transistorised ones already described. The video detector feeds TR1 common-emitter amplifier, in the emitter circuit of which is a 4·43 MHz rejector, while TR1 feeds the base of TR2 common-collector amplifier through the luminance delay line. The luminance signal in TR2 emitter circuit is developed across the contrast control, which feeds the required level of signal to the base of TR3, another common-collector stage. This in turn feeds the signal to the base of the common-emitter output transistor TR4, the collector of which feeds the luminance signal to the tube cathodes, via the drive presets in the usual way.

The Mullard BF186 output transistor has a maximum dissipation of 2·5 W with a minimum collector–emitter breakdown of 165 V, and the circuit is arranged to take full advantage of this power dissipation at a relatively high collector–emitter voltage, picked up at 285 V from the supply line. The choice of the collector load, in conjunction with the power and voltage, gives approximately 140 V to accommodate the video signal swings—this is divided into 105 V black-to-white signal for maximum tube drive, 20 V shift for the brightness control and 15 V for blanking.

Black-level is set nominally at 140 V, which means that with a typical tube cut-off of 110 V, the bias on the grids is set at 30 V. At full contrast setting, the black-to-white signal at the base of TR4 is about 3 V, which calls for a stage gain of about × 35.

It will be seen that the emitter circuit of TR1 delivers a signal from a preset control for vision a.g.c., while the collector circuit of TR2 delivers a signal suitable for working the sync separator.

Brightness control works similarly to that in a valve circuit, with the control regulating the base bias of TR3. Since TR3 is d.c. coupled to the cathodes of the picture tube, any change in bias here is reflected as a change in picture tube bias, thereby giving a change in screen illumination.

H.f. compensation is provided in the output stage collector circuit by shunt-series peaking coils. Emitter compensation is also adopted.

BLANKING AND D.C. RESTORATION

Field and line blanking is achieved by the action of TR5 in TR4 emitter circuit in the manner already described. The diode in the brightness control circuit provides d.c. restoration, which is required by the presence of the coupling capacitor, by rectifying the tips of the sync pulses and so giving a positive bias

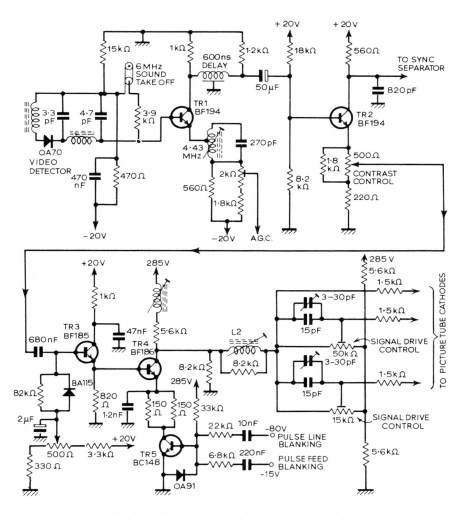

FIG. 8.11. *Transistorised luminance channel (Mullard).*

at the transistor base. A different circuit is sometimes used here, and this employs a pair of diodes which are keyed by negative- and positive-going line pulses so as to provide a black-level (d.c.) clamp on the back porches of the line sync pulses. This is shown in Fig. 8.12.

TRANSISTORISED COLOUR-DIFFERENCE AND MATRIX CHANNEL

A completely transistorised colour-difference circuit and G − Y matrix is given in Fig. 8.13. TR1 and TR3 receive at their bases respectively R − Y and B − Y signals from the red and blue detectors in the decoder. These signals are amplified by TR4 and TR6 to sufficient level to drive the red and blue grids of the picture tube.

TR2, which is in common-base configuration, picks up the correct proportions and signs of the R − Y and B − Y signals from TR1 and TR3

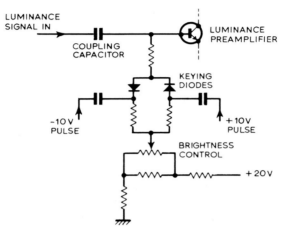

FIG. 8.12. *Black-level clamping to the back porches of the line sync pulses by two diodes keyed by ±10V line pulses.*

emitters to give the missing G − Y signal. It will be recalled the G − Y = −0·51(R − Y) − 0·19(B − Y), and this is handled in TR2, which then feeds the G − Y signal to the green output stage, TR5, for driving the tube's green grid.

DRIVE PRESETS

Green and blue drive presets allow the tube grid signals to be correctly proportioned to accommodate the variations in drive requirements for each colour. A PAL weighting factor compensation of the signals is achieved within the circuit, but final tailoring is handled by the drive presets in most circuits.

It will be seen that capacitive coupling is used to the tube grids, and because of this d.c. restoration is necessary. A double-diode in each grid feed ensures that the signals hold to a constant d.c. level during the scanning periods, and it is upon this d.c. that the a.c. signals ride.

Equal and opposite clamping pulses (about 150 V peak) are fed in through the 33 nF capacitors, and each circuit is arranged so that the diodes remain in a non-conducting state between pulses for maximum signal drive at the grids.

TRANSISTORISED INTEGRATED LUMINANCE AND COLOUR-DIFFERENCE CHANNEL

The luminance and colour-difference channels used in the BRC 2000 series all-transistor sets are rather interesting because the signals of each primary-colour are derived from a 'matrix' transistor prior to the tube guns. This means that

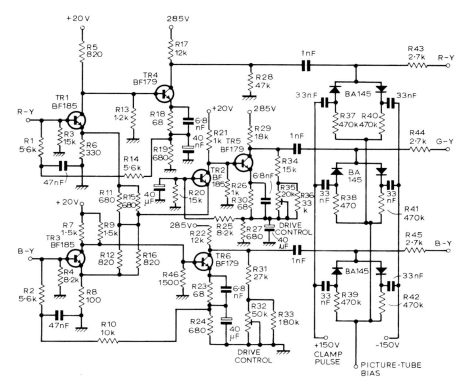

FIG. 8.13. *Transistorised colour-difference and G − Y matrix delivering red, green and blue colour-difference signals of suitable level for tube grid drive (Mullard).*

the primary red, green and blue signals can be fed straight to the *cathodes* of the picture tube—not the colour-difference signals to the grids and the Y signal to the cathodes. The complete 'video' circuit of this series of receivers is given in Fig. 8.14 (pages 148 and 149).

Luminance signal is fed to the base of TR1 and coupled from the collector and luminance delay line to the base of TR3, which is in the common-collector configuration (e.g. emitter-follower).

R − Y signals (from the red (V) detector) are fed to TR7 base, and from the collector to TR8 base. Since this transistor is in the common-collector configuration, a low impedance R − Y drive is provided to the emitter of the video transistor TR10.

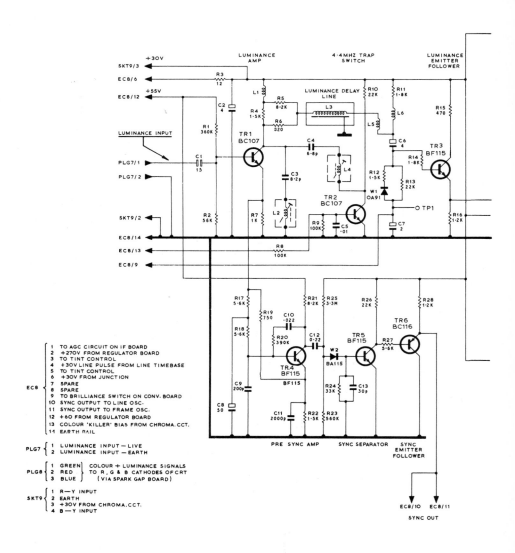

FIG. 8.14. *Transistorised video*

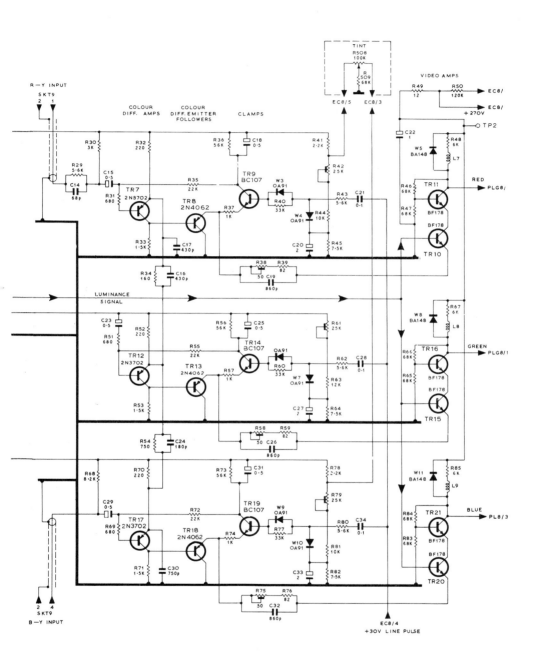

channel (BRC 2000 series).

DERIVING THE PRIMARY-COLOUR SIGNALS

An important point here is that the luminance signal from TR3 is applied to the base of TR10. Thus, TR10 acts as a matrix (like the guns of the tube in valve sets), and the red primary signal appears in its collector. This is amplified by TR11, and fed from the collector to the red cathode of the picture tube.

Exactly the same sort of thing happens to the B — Y input at TR17, and the blue primary signal appears at TR21 collector and is fed to the blue cathode of the tube.

The G — Y signal is derived by the correct proportions and signs of R — Y and B — Y signals being fed to the emitter of TR12 from the emitters of TR7 and TR17 respectively. TR13 thus serves as the G — Y driver to TR15 and, as with the other stages, luminance signal is fed to its base, thereby giving the green-primary signal at its collector, which is amplified by TR16 and fed to the green cathode of the tube.

Brightness is controlled by the three *grids* together being fed with a stabilised potential (via the beam current limiter—Chapter 7) from the brightness control potentiometer. The grids are also free to receive blanking pulses.

Each colour-difference output transistor (emitter-follower) has a clamp transistor associated with it which works in the same way as the clamp triode in

Fig. 8.15. *Video module containing the circuit sections of Fig. 8.14 (courtesy, British Radio Corporation Ltd.).*

each picture tube grid circuit of valve sets. The clamps in Fig. 8.14 are triggered by a +30 V line pulse applied to their bases. Clamp level presets are incorporated in each channel which maintain the d.c. component. Differential 'black-level' regulation over the red and blue channels is also provided, however, by the tint control, to which reference has already been made.

Another interesting feature of the circuit in Fig. 8.14 is the transistor 'switch' (TR2) controlling the 4·43 MHz colour subcarrier filter in the luminance channel. This, of course, represents the notch-filter which is only required on colour transmissions. The circuit is thus arranged automatically to switch out this filter on monochrome, thereby ensuring the maximum black-and-white definition, while leaving the 6 MHz intercarrier sound filter in circuit.

The method of operation is as follows. The base of TR2 is fed with a bias from the colour-killer bias supply which biases the transistor fully on, but only when the signal carries colour information, of course. Under this condition C4 is in series with L4—forming the 4·43 MHz filter—and are connected, via the transistor, into the luminance channel. When there is no colour, however, TR2 is switched off and the filter is automatically disconnected. Some sets use a switching diode for the filter. The 6 MHz intercarrier sound filter comprises C3 and L2 which is always in circuit.

DERIVING SYNC

Figure 8.14 also shows the sync separator section of the set. First the sync information at the emitter of TR1 is amplified by TR4. It is then fed to the sync separator proper, which is TR5, and then the output pulses are coupled to the common-collector stage TR6. From the emitter of TR6 they are delivered at the correct phase and impedance for synchronising the line and field timebase generators.

A picture of the video module is given in Fig. 8.15. The luminance delay line can be clearly seen along one side, while the groups of three bias presets and gain presets can be seen towards the middle of the board.

CHAPTER 9

Vision, Chroma, Reference Generator and Sound Stages

So FAR we have covered the ground represented on the right-hand side of Fig. 9.1. The next stage is to consider the circuits responsible for the production of the vision (monochrome and colour) and sound signals; also to see how these signals are channelled into and handled by their associated stages.

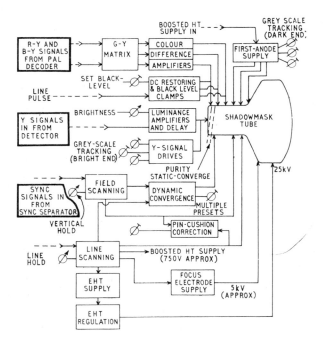

FIG. 9.1. *The 'blocks' in thin-line show the stages so far dealt with, while the thick-line 'blocks' indicate the signal sources considered in Chapters 9 and 10.*

Figure 9.2 gives a block diagram of the first stages of a typical colour set. This takes in the tuners, i.f. stages, detectors and the sync separator and a.g.c. and intercarrier-sound stages. The dual-standard functions are virtually the

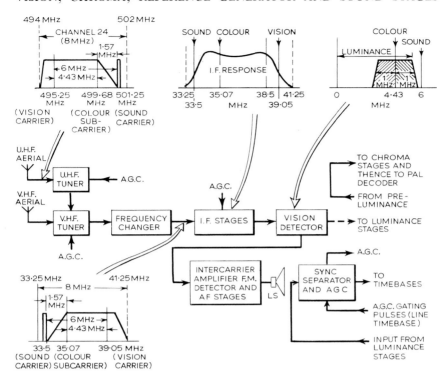

FIG. 9.2. *Block diagram of the first stages of a typical receiver, showing signal characteristics, frequency placements and circuit feeds to the various points.*

same as in monochrome early stages and are fully dealt with in *Television Servicing Handbook*. There are extra complications in the line and dynamic convergence sections but these are dealt with in Chapter 7.

TRANSISTOR FRONT-END

Practically all British colour sets have transistor early stages—the tuner, i.f. and intercarrier sections. There is also a trend for transistors to be used for an integrated tuner, covering all v.h.f. and u.h.f. channels. Moreover, single-standard sets incorporate a u.h.f. tuner only (most models, anyway) and this always uses transistors. The inside of such a tuner, using quarter-wave lines, is shown in Fig. 9.3(*b*) on page 156.

Transistors in the tuner, especially silicon ones, ensure the best possible signal-to-noise performance, which is particularly important for good colour reproduction. The decoding and encoding processes have a detracting influence on the subjective noise performance, relative to monochrome, which is one of the reasons why more signal is needed by colour sets for a given noise performance.

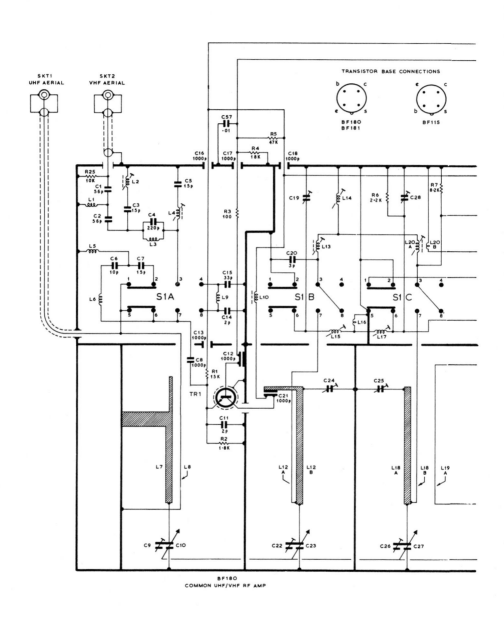

FIG. 9.3. *(a) Circuit diagram of transistorised integrat*

154

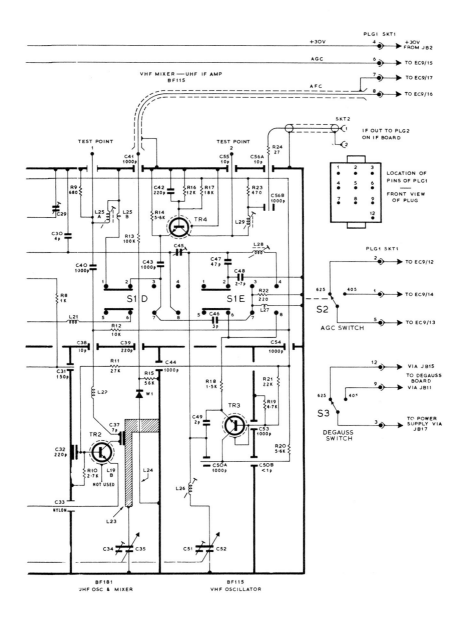

uner, incorporating facilities for a.f.c. (BRC).

INTEGRATED TRANSISTORISED TUNER

The circuit of the BRC integrated tuner is given in Fig. 9.3(*a*). This uses quarter-wave tuning lines for the u.h.f. bands, tuned over the channels in the usual way by a four-gang capacitor (C10, C23, C27 and C35). The v.h.f. channels are

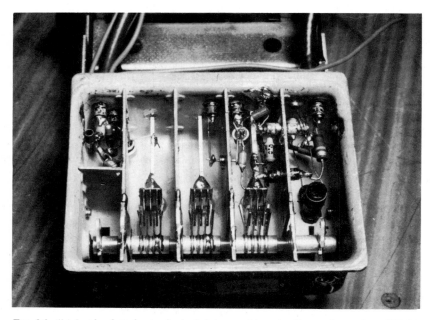

Fig. 9.3. *(b) Inside of single-standard u.h.f. tuner, with tuning lines and capacitors (BRC).*

also tuned by this gang, but in conjunction with the more conventional type of inductors.

TR1 is the r.f. amplifier transistor for u.h.f. and v.h.f., TR2 is the u.h.f. local oscillator and mixer, TR4 is the v.h.f. mixer which also acts as an extra i.f. amplifier on the u.h.f. channel, and TR3 is the separate v.h.f. oscillator. It will be seen that the v.h.f. oscillator is tuned by a *fifth* section on the tuning gang.

The general design of the tuner is fairly conventional, and it follows the accepted v.h.f. and u.h.f. tuner techniques. In spite of the v.h.f./u.h.f. integration, however, there are two aerial inputs, one for u.h.f. and the other for v.h.f. This is really desirable, since it avoids having to use a single coaxial downlead for both signals, and thus makes it possible to engineer the u.h.f. aerial system separately for maximum efficiency and the least loss in the downlead.

It will be seen that the tuner operates an a.g.c. switch (S2) which changes the control between the two standards, also a degauss switch (S3), which energises the automatic degaussing system each time the tuner is switched to a channel of different line standard.

A feature which is becoming common at u.h.f. is automatic frequency correction (a.f.c.). This is operated by the capacitor-diode W1 coupled by an inductive loop (L24) to the u.h.f. oscillator tuning line. When such a diode is reverse biased its capacitance bears a relationship to the degree of bias, and its value reduces as the bias is increased.

Since the loop is connected in parallel with this capacitance, the effect that it has on the tuned oscillator frequency depends on its value. This means that the tuned frequency can be altered by changes in the value of the capacitance, which which simply means adjusting the reverse bias applied to the diode.

The bias is obtained from the a.f.c. circuit in the i.f. channel (Fig. 9.11), and is then tapped from the potential divider (in the tuner) R13, R15. Thus, any change in potential here will automatically change the oscillator frequency. The bias circuit delivers a potential of suitable polarity to restore the correct tuning whenever there is any tendency for the u.h.f. oscillator to drift.

Because it is very important to maintain absolutely the correct tuning on colour sets for the optimum reproduction of colours, a.f.c. is a highly useful embodiment, and is even more important on those sets using press-button channel selection, where the depressed button itself needs to be rotated to establish the correct tuning point. Without a.f.c., progressive detuning can occur due to the normal action of the push-button, and not all viewers are inclined to correct the tuning themselves each time the programme is selected.

Figure 9.2 demonstrates that the luminance channel, from the tuner to the vision detector, is the same as in a monochrome set. A.G.C. is fed to the tuner and i.f. strip in the usual manner, and the exact nature of this control is sometimes switched between standards, as described in Chapter 8 (also later in this chapter).

The sync separator picks up its signal from the luminance output as it does in a monochrome set at the video output valve, and this section may incorporate the a.g.c. bias circuit, gated on 405-line standards from the line timebase.

In Fig. 9.2 the chroma signal and the bursts are extracted from the main vision detector, but this is not a universal arrangement. For instance, Fig. 9.4 shows that in the Philips G6 chassis the i.f. channel is split after the second stage

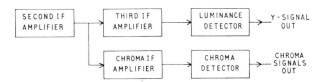

FIG. 9.4. *A separate chroma detector is sometimes used.*

to give a separate chroma i.f. channel. The chroma i.f. is then fed into its own chroma detector. The Bang and Olufsen 3000 series uses a similar arrangement with the chroma detector also delivering the intercarrier signal. The BRC 3000 series uses separate luminance and chroma detectors, with the latter also delivering intercarrier signal, but both are fed from a common final i.f. stage. When the intercarrier signal is taken from a detector other than that delivering the luminance signal, it is possible to include a 33·5 MHz rejector in the luminance detector circuit to reduce significantly the effect of the beat between the chroma and sound signals. Such a rejector is used in both the Bang and Olufsen 3000 series and the BRC 3000 series, as well as in other models.

The frequency distribution characteristics of the signal within the 625-line standard 8 MHz channel are shown at various points in Fig. 9.2. The disposition of the component signals in the spectrum is shown at the aerial input. The spacings apply to all the u.h.f. channels, of course, not merely to Channel 24 illustrated.

The luminance signal is 'dovetailed' (see Chapter 5) with the chroma signal consisting of the $B - Y$ and $R - Y$ or U and V signals amplitude-modulating a subcarrier which is itself suppressed. Double-sideband modulation is used, but the modulation extends only to 1 MHz. The subcarrier frequency is 4·43 MHz—to be exact 4·43361875 MHz, derived as explained earlier.

The spectrum at the i.f. input is the signal after the frequency-changer stage. Of course the carrier spacings remain:

(*a*) sound and vision carriers 6 MHz

(*b*) sound and colour subcarrier 1·57 MHz

(*c*) colour subcarrier and vision carrier 4·43 MHz.

The next response curve shows how these signals are handled in the i.f. channel. The response has to be adequate for the colour subcarrier, but it is down slightly at the vision carrier (single sideband working) and well down at the sound carrier.

INTERCARRIER BUZZ

We are dealing here solely with reception of the 625 standard, using intercarrier sound. We know from monochrome practice that excessive response at the sound carrier frequency can incite intercarrier buzz and picture shortcomings, while too little response can cause weak, noisy sound.

Things are even more critical for colour because the colour subcarrier falls only 1·57 MHz from the sound carrier, which means that excessive response at the sound frequency can generate 1·5 MHz patterns on the picture, resulting in herringbone interference, as well as the intercarrier buzz.

COLOUR NOISE

Too great a response at the colour subcarrier frequency (the subcarrier falling on a response peak) can exaggerate saturation. Too small a response will weaken the colour and cause colour noise, sometimes referred to as the 'rainbow' or 'confetti effect'.

This is one reason why the aerial signal should be as strong as possible, and the aerial response flat over the entire channel width. Normally, of course, u.h.f aerials embrace a whole group of channels. The same applies to any aerial amplifier or booster and to the u.h.f. diplexer, if used. The complete colour picture is composed of signals in two primary channels, so to speak (luminance and chroma), and it is important for the aerial signal to be sufficiently strong to maintain at least a 50 dB signal-to-noise ratio in the chroma channel as well as in the luminance channel.

IMPORTANCE OF CORRECT ALIGNMENT

The importance of correct i.f. alignment is thus revealed, and the design of these circuits has to ensure that the response characteristics are maintained over the range of vision a.g.c. and contrast control.

I.F. CHANNEL REJECTORS

The response has to be carefully 'tailored' at the high and low limits to avoid beats from the carriers of adjacent channel transmissions. Fortunately, u.h.f. aerials are or should be designed for optimum protection against adjacent channel and co-channel stations, but rejectors are still required in the i.f. stages to prevent possible intrusion of adjacent carriers during abnormal reception conditions.

On the 625-line standard the adjacent sound rejector is tuned to 41·5 MHz (affording protection against Channel 1 sound breakthrough too) and the adjacent vision rejector to 31·5 MHz. A rejector is focused on the in-channel 625-line sound i.f., giving a dip of about 20 dB at 33·5 MHz, and it is this that sets the level of the sound carrier in the i.f. channel.

Dual-standard i.f. stages have further rejectors switched in on the 405-line standard. These reduce the response bandwidth and also attenuate the 405-line adjacent carriers. The technique here, though, is virtually the same as with dual-standard monochrome sets.

The i.f. stages feed the detector in the usual way and this delivers a video signal disposed over a spectrum as shown in Fig. 9.2. Luminance information ranges from zero (d.c.) to about 5·5 MHz and within this the sidebands of the R — Y and B — Y signals occupy ±1 MHz with the lower sideband starting at about 3 MHz (Fig. 9.5). Between the sidebands the 4·43 MHz subcarrier is located and, finally, there is the intercarrier sound signal at 6 MHz.

The detector also gives sync pulses and 'colour bursts'. The latter, located on the back porches of the line sync pulses, 'tell' the receiver the exact frequency

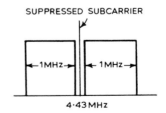

FIG. 9.5. *Chroma sidebands occupy a spectrum of ±1MHz, centred on the subcarrier frequency (4·43MHz).*

and phase of the suppressed colour subcarrier. The detector also delivers a signal at 1·57 MHz as the result of the sound carrier beating with the colour subcarrier. If the strength of this is excessive (as could result from misalignment of the sound carrier rejector in the i.f. channel) it causes herringbone patterns, as already mentioned.

NOTCH FILTER

The luminance or Y signal is coupled from the detector to the luminance pre-amplifier stages which are considered in Chapter 8. A 4·43 MHz rejector is included in the luminance path which puts a fairly substantial dip in the video response at the colour subcarrier frequency for the purpose of minimising 4·43 MHz dot-pattern on monochrome. (A similar effect occurs in monochrome sets on the 405-line standard due to the 3·5 MHz beat between the vision and sound carriers, and many sets of course have 3·5 MHz dot-pattern rejectors in the video output stage cathode circuit.)

Excessive colour subcarrier can also result, on arriving at the cathodes of the shadowmask tube, in a desaturating effect on the colour. This is caused by a rectifying action arising from the non-linear characteristic of the tube.

Figure 9.6 shows a 4·43 MHz filter with its response characteristic. This is an important filter in colour sets and must be adjusted very accurately to the sub-carrier frequency for maximum interference rejection. All sets have a filter of

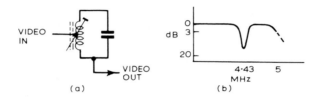

FIG. 9.6. *Notch filter used in the luminance channel (a) and its response curve (b).*

this kind, called a *notch filter*, but sometimes a low-pass filter is included in addition. This has the effect of progressively decreasing the video response from about 4 MHz (or sooner) so that reasonable attenuation occurs at the subcarrier frequency. As would be expected, this latter kind of rejection, unless well adjusted, can cut down the picture detail.

It will be recalled that some sets incorporate an 'electronic switch' (transistor stage) which brings in the notch filter only when the colour signal is being re-ceived, thereby ensuring that the filter does not degrade the monochrome reception.

While on the subject of chroma interference, it is worth noting that the effect is also present to some extent when a black-and-white set is receiving a colour transmission—depending on how well it can resolve detail—and is most troublesome in picture areas corresponding to intense colour. Excessive colour subcarrier, as on highly saturated picture areas, can also overcontrast the monochrome display a little for the reason just mentioned for colour tubes.

The colour-bursts can also produce symptoms on monochrome sets similar to those of slight line retrace corona; that is vertical lines of dots (or dashes) down the left of the screen. This happens because the burst occurs during the later part of the line retrace and this stretches out the resulting dots so that they appear under the start of the line scan.

To return to the main theme, Fig. 9.7 shows a video detector circuit and the intercarrier take-off across a transformer tuned to 6 MHz. This has a rejector effect on intercarrier signal as far as the luminance feed is concerned.

The composite video signal across the detector load is coupled directly to the luminance preamplifier and the chroma information is extracted after the first luminance stage. This is not a universal rule, but is a neat method of handling the chroma take-off since it has little loading effect on the detector proper. In

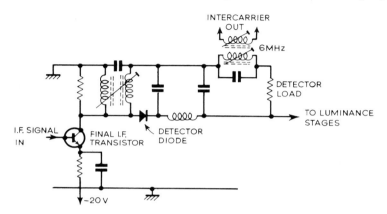

FIG. 9.7. *Method of deriving the intercarrier sound signal from the 625-line main detector.*

such circuits the sync separator is coupled-in after the second luminance stage, from whence the luminance signal itself is taken to the output stage.

In the Philips G6 chassis, with separate luminance and chroma detectors, the luminance detector feeds into a preamplifier stage whose collector is loaded into the intercarrier take-off transformer, as shown in Fig. 9.8. This stage acts as a phase-splitter, delivering 625-line phased signals from its collector (following the intercarrier take-off) and 405-line phased signals from its emitter. The a.g.c. amplifier is also fed from the collector. Output is channelled through a 3·5 MHz sound/vision dot-pattern filter on the 405-line standard and through a 4·43 MHz colour subcarrier filter on the 625-line standard.

The luminance delay line, which in different sets ranges from about 600 ns to 830 ns, depending on the relative bandwidths of the luminance and chroma channels, is also in the collector circuit, and is in series with the 625-line output.

The chroma detector (Fig. 9.9), is fed from a separate chroma i.f. stage which picks up signals from the collector of the second main i.f. stage. The detector's output is loaded on one side into the chroma bandpass filter and the chroma amplifier input filter and, on the other side (with separate diode), into a tuning meter.

Another variation is seen in the Decca CTV25 chassis. This also uses two detectors (Fig. 9.10), but one is for the intercarrier signal only. The video detector feeds into a phase-splitter for luminance and chroma signals. This enables the video detector to carry a 33·5 MHz rejector for chroma/sound beat elimination.

In all sets the chroma signal at the video detector (or chroma detector if separate) is fed to the chroma amplifier through a bandpass filter which keeps out the luminance signals from d.c. to about 3 MHz. Often intercarrier and

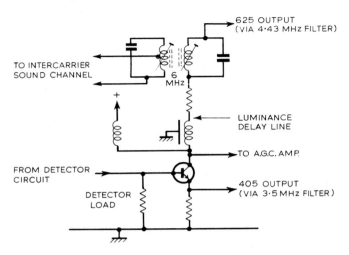

FIG. 9.8. *Simplified circuit showing intercarrier 'take-off' in the luminance amplifier collector circuit.*

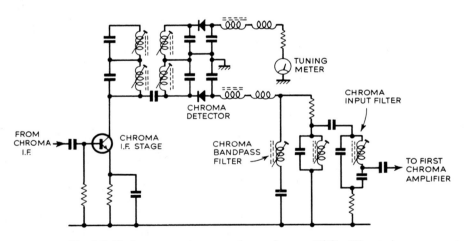

FIG. 9.9. *Tuning meter and separate chroma detector (Philips G6 series).*

sound/colour-subcarrier rejectors at 6 MHz and 33·5 MHz respectively are included as well. The idea is to clear the chroma input of all unwanted signals.

The circuit of the BRC all-transistor i.f. section is given in Fig. 9.11. There are three stages in the vision i.f. amplifier using TR1, TR2 and TR3, all npn transistors arranged in the common-emitter mode. The signal from the i.f. output

162

of the tuner (Fig. 9.3a) is fed to the base of TR1 through one of two filter circuits, depending on the setting of the standard switch S1A/B. The selected filters give the i.f. channel the required response characteristics, as already discussed.

The detector diode for the 625-line standard is W1 and for the 405-line standard W3. These are arranged, of course, to correspond to negative- and positive-going vision modulation respectively, in the usual way.

The luminance output from the selected standard is obtained from switch S1G, while S1D selects the appropriate contrast control. The luminance output is fed to the video module (Fig. 9.14).

Chroma signal is fed from the 405-line standard detector, on 625 lines, via C49, to the base of the chroma emitter-follower, which is TR11 wired in the common-collector mode, the chroma signal thus being delivered at the emitter (and then passed to the chroma amplifier stages). The chroma signal at TR11 input is 'tuned' by L40 and C107/108. This stage, incidentally, is biased by the

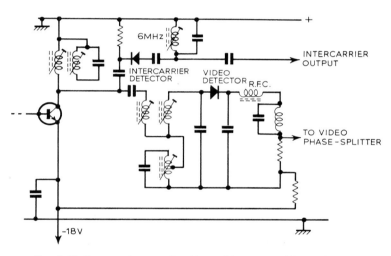

FIG. 9.10. *Separate detectors for video and intercarrier (Decca CTV25).*

automatic chroma control (a.c.c.) potential derived from the burst signal (see Chapter 10, in particular). Another stage in the chroma channel proper is muted in the absence of chroma signal but biased-on by the ripple signal. This is the 'colour killer' stage (see page 172).

Intercarrier sound signal at 6 MHz is picked up from the transformer L17/18 in the 405-line standard vision detector circuit at the junction of C42/43, and is fed to the sound i.f. stages TR4 and TR5, which are also npn transistors in the common-emitter mode.

These stages, too, are designed to respond to the ordinary 405-line standard sound i.f. signal by reason of the dual tuning in their base and collector circuits. The 6 MHz 625-line standard sound signal, which is frequency modulated, of course, is fed to the ratio detector comprising W5 and W6. The audio output is developed across C82 and fed to the audio channel through S1E and C88.

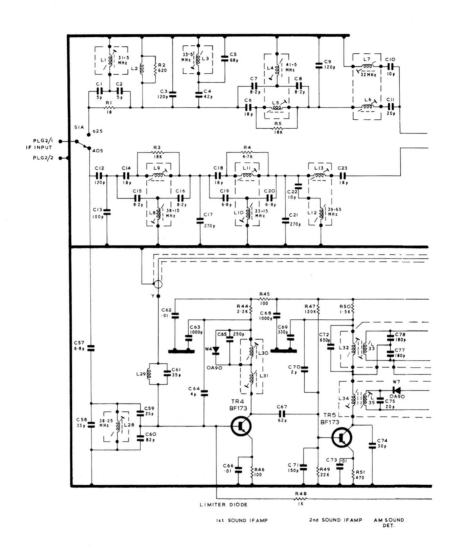

FIG. 9.11. *I.F. and detector stages (BRC 2000 series)*

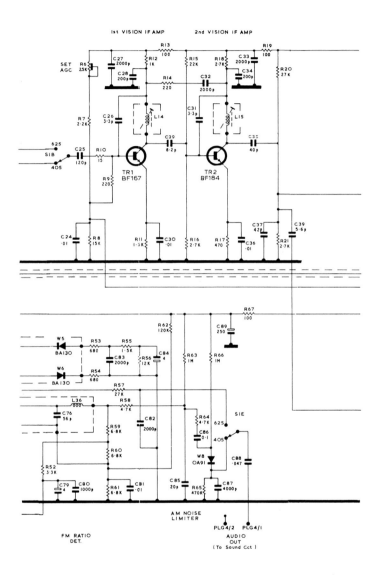

The circuit is continued on pages 166 and 167.

165

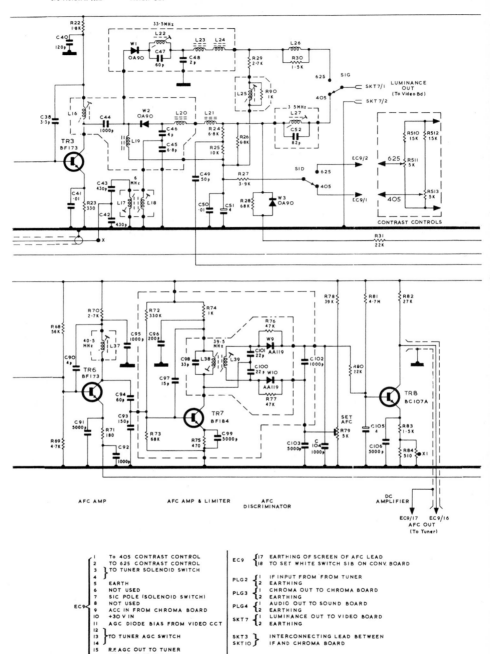

3rd VISION IF AMP   VISION DET

AFC AMP   AFC AMP & LIMITER   AFC DISCRIMINATOR   DC AMPLIFIER

EC9/17 EC9/16
AFC OUT
(To Tuner)

EC9
1 To 405 CONTRAST CONTROL
2 TO 625 CONTRAST CONTROL
3 TO TUNER SOLENOID SWITCH
4
5 EARTH
6 NOT USED
7 SIC POLE (SOLENOID SWITCH)
8 NOT USED
9 ACC IN FROM CHROMA BOARD
10 +30 V IN
11 AGC DIODE BIAS FROM VIDEO CCT
12
13 TO TUNER AGC SWITCH
14
15 R.F. AGC OUT TO TUNER
16 AFC OUT TO TUNER

EC9 { 17 EARTHING OF SCREEN OF AFC LEAD
       { 18 TO SET WHITE SWITCH SIB ON CONV. BOARD

PLG2 { 1 IF INPUT FROM FROM TUNER
       { 2 EARTHING

PLG3 { 1 CHROMA OUT TO CHROMA BOARD
       { 2 EARTHING

PLG4 { 1 AUDIO OUT TO SOUND BOARD
       { 2 EARTHING

SKT7 { 1 LUMINANCE OUT TO VIDEO BOARD
       { 2 EARTHING

SKT3 } INTERCONNECTING LEAD BETWEEN
SKT10 } IF AND CHROMA BOARD

FIG. 9.11 (cont.). *F. stages, detectors, chroma take-off, a.f.c. an*

166

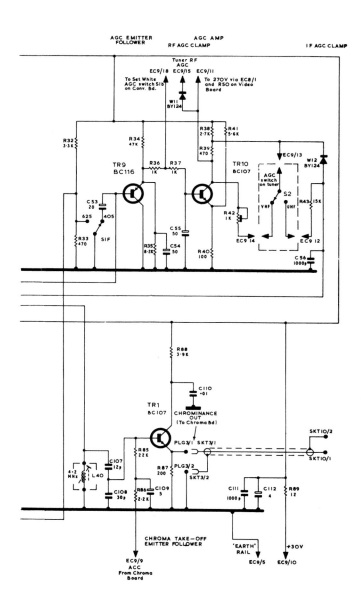

a.g.c. circuits (BRC 2000 series). TR1 above should be labelled TR11.

The 39 MHz 405-line standard i.f. signal is fed to the a.m. detector diode W7, and the resulting audio signal is fed to the other pole of S1E, via the a.m. noise limiter, using diode W8. These are fairly conventional circuits.

A.F.C. POTENTIAL

Signal at the base of the third vision i.f. amplifier TR3 is also fed, via C39, to the base of TR6, which is the a.f.c. amplifier. This signal is further amplified and limited in amplitude (clipped) by TR7, and then developed across the discriminator transformer L38–L39. L39 secondary feeds the discriminator diodes W9 and W10, and the action of this circuit is such that provided the vision i.f. signal corresponds exactly to the tuning of L38–L39 both diodes conduct equally, and zero potential is developed by the circuit across C102. However, should the i.f. signal tend to drift a little from the tuned frequency of the transformer windings, a positive or negative voltage arises at the top of C102 with respect to its bottom connection, depending on whether the drift is up or down in frequency. This happens, of course, when the oscillator frequency of the tuner drifts.

This potential is amplified by TR8 (which is a common-emitter d.c. amplifier) and fed to the tuner capacitor and diode (Fig. 9.3a) from TR8 collector, thereby constituting the potential needed by the a.f.c. system.

The preset potentiometer, R79, allows the bias on the capacitor-diode to be set for the correct plus and minus range of auto control, for this simply feeds a standing bias from the supply line to the diode.

A.G.C. POTENTIAL

The final stages in Fig. 9.11 are those concerned with the a.g.c. potential, and comprise TR9 and TR10. The first is an emitter-follower which receives a video signal at its base from the vision detectors, via R28, W3 and R31. This input is referred to black-level and then amplified by TR10, at the collector circuit of which appears a d.c. potential of a magnitude dependent on the strength of the vision signal. This, of course, is the a.g.c. bias, which is applied, via clamps, to the tuner and vision i.f. transistors.

W11 is the tuner a.g.c. clamp diode, which is biased from the 270 V supply line via a resistor. This means that a.g.c. bias is applied to the tuner transistors only when the vision signal exceeds a predetermined level, thereby ensuring the best signal-to-noise performance.

The i.f. channel a.g.c. bias is clamped by W12, and fed negatively from its 'anode' to the base circuit of TR1. There is a preset control in this circuit (R6) to establish the correct range of auto control.

It will be seen that the a.g.c. bias is modified between the two standards, by the action of switch S2, located on the tuner. Switch S1F also modifies the a.g.c. input to TR9 and mutes the a.f.c. amplifier on the 625-line standard, on which a.f.c. is not employed. A photograph of the i.f. module is given in Fig. 9.12.

TRANSISTORISED SOUND CHANNEL

The sound channel circuit of the BRC dual-standard sets is given in Fig. 9.13. Little need be said about this as it follows the conventional practice of most such transistorised stages. The common-emitter voltage amplifier npn transistor TR8 receives the detector signal from the sound channel (Fig. 9.11), via the volume control R505, and passes this on from its collector to the base of the second amplifier TR9.

Both TR10 and TR11 are fed at their bases from TR9 collector circuit to give a push–pull drive to the output transistors TR12 and TR13, which are biased for class B working. This is possible because TR10 and TR11 are complementary

FIG. 9.12. *BRC 2000 series i.f. module (courtesy, British Radio Corporation Ltd.)*.

—TR10 is npn and TR11 is pnp. The correct drive signals for the output transistors are obtained for TR12 from TR10 emitter and for TR13 from TR11 collector, and the loudspeaker is coupled to the emitter–collector series circuit of TR12 and TR13 through the electrolytic capacitor C25.

Stabilising feedback is provided from the output pair back to TR9 base through R45, which means that TR9 collector–emitter current affects the output pair balancing, and this is adjusted by the preset R49.

Some colour sets still employ valve operated audio stages, using a triode pentode or a push–pull circuit with negative feedback for improved quality, but the general trend is towards transistorised audio stages.

To sum up, the detector delivers the luminance signals, the chrominance signals, and the colour bursts (Fig. 9.14). The luminance signals are filtered into the luminance amplifier and thence sent to the cathodes of the shadowmask tube. We have already seen how this is done, which leaves the chroma and colour-burst signals to be dealt with. The chrominance (chroma) section is that between the vision or chroma detector on the PAL decoder.

Block diagram Fig. 9.15 shows the stages in the chroma section proper. These are mostly amplifier stages, transistor or valve, and they feed the sidebands of the R − Y and B − Y signals to the detectors. From the detectors come the colour-difference signals for the tube grids.

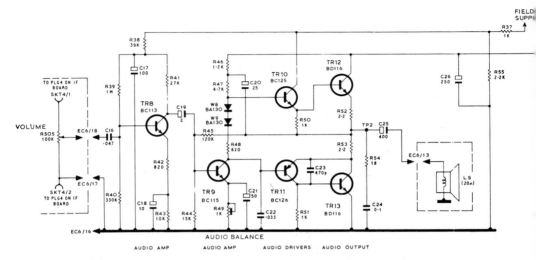

FIG. 9.13. *Transistorised audio section (part of the field timebase module of the BRC 2000-series).*

This decoding action calls for a phase-controlled reference signal. This is obtained from a reference oscillator which is synchronised to the colour sub-carrier at the transmitter by means of the colour-bursts.

Figure 9.16 gives a block diagram of stages from the burst take-off (see Fig. 9.15) to the detectors—taking in other rather important sections on the way. Now let us see how these stages operate, starting with those in Fig. 9.15.

FIG. 9.14. *The detector delivers chroma signals to the PAL decoder, colour bursts to lock to reference generator and luminance signal. The PAL decoder output is in the form of colour-difference signals.*

It is important that the chroma amplifier receives only chroma and colour-burst signals from the detector, and this is ensured by a high-pass filter at the input. The filter in the Philips G6 chassis is shown in Fig. 9.9. It is designed to attenuate the d.c.–3 MHz part of the luminance signal. The luminance signal itself is separately filtered into the luminance amplifier. Other filter sections (as

170

already discussed) are sometimes incorporated for the removal of the 1·57 MHz sound/subcarrier beat and the 6 MHz intercarrier sound signal.

The first chroma amplifier stage thus operates over about 3–6 MHz, with the sound intercarrier notched out. The signal consists of the chroma sidebands (centred on 4·43 MHz) and the colour burst at 4·43 MHz.

The aim is to get the chroma signal and a reference signal at subcarrier frequency into the PAL decoder, which will then yield the R − Y and B − Y signals.

The colour bursts are used to keep the locally generated reference subcarrier in correct frequency and phase. They are generated on the back porches of the line sync pulses, diverted from the chroma amplifier and then processed separately, as shown in Fig. 9.16.

The chroma passband of ±1 MHz is achieved by bandpass couplings. Most sets have two, or perhaps three, stages of chroma amplification and this is subjected to manual and auto controls, as in Fig. 9.15.

COLOUR CONTROL AND A.C.C.

It is not uncommon for the gain of an early stage in the chroma channel to be controlled automatically, based on the level of the burst signal. This is called

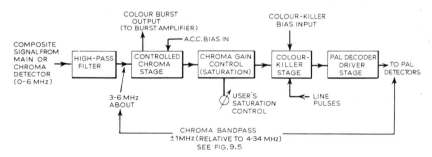

FIG. 9.15. *Stages of the chroma section up to the PAL decoder.*

either chroma a.g.c. or automatic colour (or chroma) control—a.c.c. for short. Most sets also have a manual control for chroma channel gain. This is the chroma equivalent of the contrast control, and is called the *saturation control* or *colour control*.

The gain of the chroma channel determines the level of the multiplexed colour-difference sidebands (e.g. chroma signal) applied to the PAL decoder. On turning the control right back the chroma signal falls to zero and colour fades from the picture. The luminance stages are not affected, of course, so the monochrome representation remains.

Not all models have a user's colour control. Some rely on a.c.c. and a colour preset control. Others have the colour control electrically interlocked with the contrast control, as we have already seen. Some models without a saturation control feature a user's tint control for *colour balance*, which enables the viewer to select the flesh tones most preferred either direct from a colour picture or from the picture inside the circle of Test Card F which has been specially designed for this purpose.

COLOUR KILLER

Another auto control in the chroma channel is the 'colour killer'. This is an automatic 'switch' that operates from the colour transmission. The 'switch' generally takes the form of a stage in the chroma channel which is biased to signal cut-off but which automatically 'biases-on' when colour bursts are present.

The last but one block in Fig. 9.15 shows the colour-killer stage and also that a bias is applied to control the switching. The bottom block in Fig. 9.16 shows from whence this bias might be obtained. It will also be seen that there is an alternative input to the colour-killer rectifier, from the burst amplifier, shown in broken line.

All that is required to make the chroma channel active is a bias to outweigh the cut-off bias normally on one of its stages. This can be obtained directly from

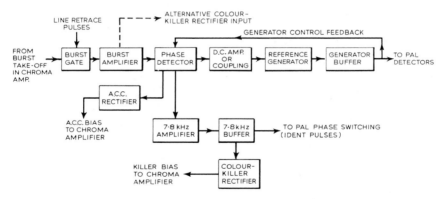

Fig. 9.16. *How the reference generator is phase controlled; also other processes required prior to the PAL decoder. Colour bursts across the phase detector transformer are rectified for a.c.c.*

the colour bursts, via a simple rectifier, or from a half line-frequency ripple (7·8 kHz) which develops in the discriminator (phase detector) controlling the reference generator, again via a rectifier.

To sum up, the bias for making the chroma channel *active* is derived from the bursts in some way or other, so that on monochrome transmissions the chroma channel is non-responsive. At first sight it would seem that the monochrome signal would not have any influence on the colour channel. But just imagine what would happen if the chroma channel were 'open' on a monochrome transmission during a spell of bad r.f. interference—such as co-channel interference which is bad in certain areas during the spring and summer months.

The unwanted r.f. might well fall within the chroma passband and the chroma channel would then activate on the interference and consequently produce a nice rainbow display on the monochrome picture! Moreover, because the chroma channel would be running 'flat out'—no a.c.c. bias—strong noise signals would be filtered to the grids of the shadowmask tube, and grain in colour would mar the monochrome rendering.

Clearly, it is essential for the chroma channel to be 'killed' when the set is working on a monochrome signal. Actually, as we have seen, the action is the opposite because the chroma channel is *biased-on* by the colour transmission. The term 'colour activator' would appear to be more appropriate.

Figure 9.17 gives the circuit of the chroma channel in the Decca CTV25. Here TR1 receives at its base the chroma signals, the input filter having removed the luminance information, and communicates these to the base of TR2, via the bandpass coupling and saturation control.

Colour bursts are taken from the collector of TR1 by the 4·43 MHz tuned circuit and are fed to the burst amplifier (see Fig. 9.19). TR2 is the colour-killer

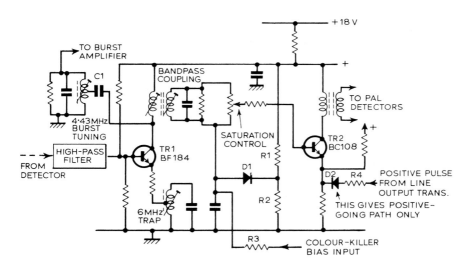

Fig. 9.17. *Chroma stages (Decca CTV25)*.

stage. This is held at cut-off by the nature of the biasing in the base circuit. When the transmission contains colour information a countering bias is fed to the base.

R1 and R2 form the base potential divider for TR2. The potential at this junction is a little positive and this is required for the transistor base because it is an npn type. Diode D1, however, is in the way of this potential and the base (via the saturation control). The base current does not flow because D1 is in reverse bias—positive potential to 'cathode'. The stage is thus cut-off.

Now, when the set is receiving a colour signal, a positive bias is fed to the 'anode' of D1, causing it to conduct in the forward direction. This then allows the potential at R1/R2 junction to reach TR2 base, which biases-on the stage and results in chroma output.

The saturation control works in similar way to an ordinary volume control.

The required level of chroma signal across its resistive element is tapped-off by the slider and fed to TR2 base.

The block diagram in Fig. 9.15 shows that line pulses are applied to the colour-killer stage, and in Fig. 9.17 these are shown going into TR2 emitter circuit, via R4 and D2. What is the purpose of this? It is not very complicated and is simply an artifice to blank the chroma channel during the line retrace period to prevent colour bursts and line sync pulses getting into the PAL decoder, where their presence could cause colour interference troubles. A neat way of blanking is by using line flyback pulses, as shown in Fig. 9.17. Here positive-going pulses push TR2 hard into signal cut-off. This is equivalent to negative-going pulses being applied to the base. The diode in series ensures that only positive-going pulses reach the emitter. The 6 MHz intercarrier rejector is located at the emitter of TR1, and is very similar to the sound rejector found in vision i.f. channels.

In practice, all the tuned circuits leading the PAL decoder are critical, especially the bandpass transformer coupling. Misalignment results in sideband cutting, phase shifts and interference patterns.

AUTO COLOUR CONTROL

In the Mullard-designed colour set the first chroma stage—common to colour bursts and chroma signal—is gain-controlled. A.C.C. potential, derived from the rectified burst signal, is applied to the base network of the controlled transistor amplifier. The specification for this circuit states that the chroma output signal is held constant within 2·5 dB for 12 dB variations in input signal. A.C.C. is often desirable to combat saturation changes due to signal fading, changes in setting of the contrast control and alterations to the fine tuning control. Of course, the composite signal in the i.f. channel is subjected to ordinary a.g.c., and some designs rely essentially on this for both monochrome and colour.

Some sets rectify the colour bursts present at the phase detector or burst gate and direct the resulting d.c. separately to the controlled chroma stage and to the colour-killer stage.

Figure 9.16 shows the a.c.c. rectifier receiving burst signal from the phase detector transformer. The same signals could, as mentioned, also be applied to the colour-killer rectifier. Alternatively, both a.c.c. and killer bias signals could be picked up from a burst-amplifier stage. It is only the PAL system that can use the 7.8 kHz ripple signals from the phase detector to drive the colour-killer rectifier, as shown in Fig. 9.16—no such ripple exists in the NTSC system.

IDENT PULSES

A 7·8 kHz signal is required to 'steer' the PAL switching—operated by line pulses—on to the correct line. These are called identification ('ident' for short) pulses, and need to be derived from a 7·8 kHz amplifier, as shown in Fig. 9.16. It is quite a simple matter, therefore, to arrange to rectify these pulses for killer bias, and many sets adopt this idea.

The chroma stages of the Philips G6 colour chassis are shown in Fig. 9.18. This is one of the few sets that uses valves instead of transistors in the chroma channel.

V1 control grid receives chroma signals via the usual high-pass filter, amplifies them and passes them to V2 control grid, via the bandpass coupling transformer.

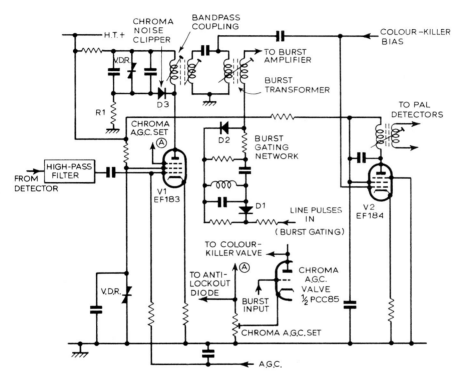

FIG. 9.18. *Chroma stages (Philips G6 series)*.

The secondary of this also couples to the primary of the 'burst transformer', tuned to extract only the burst signals, and the secondary feeds them to the burst amplifier (not shown).

An interesting feature is the method of 'burst gating' used in this set. This is an action which is necessary to ensure that the burst amplifier receives signals *only* during the time that colour bursts are present in the transmission. During the picture part of the signal the burst gate must be closed so that picture signal components do not get into the phase detector and reference generator circuits.

Figure 9.16 shows that the burst signals from the chroma channel are applied to the burst amplifier through the burst gate. In actual fact, the gate may be a function of the amplifier itself but, in all cases, the gate is operated by line-frequency pulses from some source. The effect is that the gate is opened by each pulse and is closed immediately after the line retrace period. We shall see later how this is achieved in the burst amplifier by pulse switching, but for the moment let us consider the circuit in Fig. 9.18.

Here the burst transformer secondary is gated by line pulses applied through a pulse shaping and switching network, called the burst gating network, and identified in the circuit. The action is as follows: D1 conducts on the line pulses, giving negative pulses on D2 'cathode', which causes this diode to conduct and as a result takes the bottom end of the burst transformer secondary to chassis. Until D2 conducts, there is no return circuit for the transformer secondary and hence no output. The burst gate timing is critical and hence the values of the components in these circuits.

DIODE DAMPING

Another interesting feature is the 'noise clipper' operated by diode D3. Under quiescent conditions this diode is biased non-conducting by the h.t. at V1 anode applied through R1. This condition remains right up to the full chroma signal levels but during periods of interference the peaks reverse the diode bias, causing it to conduct, and thereby attenuating the interference by the diode 'damping' the anode circuit.

Colour killing operates on V2. In the absence of colour bursts the control grid is heavily negative. When bursts are present on the signal the bias is countered and chroma signals are passed to the PAL detectors.

The circuit employs several voltage-dependent resistors for voltage control and stabilisation. It will be recalled that this kind of resistor has a high value when the voltage across it is low and a low value when the voltage is high. The component also works as a pulse rectifier.

A.C.C. bias, picked up from the burst signal, rectified and amplified, is applied to V1 control grid. The bias goes less negative as the signal falls, thereby restoring the level at the output by increasing the channel gain—ordinary a.g.c. action, in fact.

The triode circuit inset in Fig. 9.18 shows the a.c.c. amplifier and the 'set a.c.c.' control. This works in conjunction with an anti-lockout diode feeding back to the suppressor grid of V1, and prevents the a.c.c. system from being 'locked out', as used to happen on some early monochrome vision a.g.c. systems!

Figure 9.19 shows a typical burst signal channel, similar to that illustrated in Fig. 9.16 block diagram. Bursts are applied to the base of TR1, which acts both as a gate and burst amplifier. Gating is provided by pulses at line frequency fed to the base through R1. These are positive-going and switch the transistor on only during the burst periods. The bursts are developed across the primary of the phase detector, or discriminator transformer, and coupled to the diodes D1 and D2 from the secondary windings.

PHASE DETECTOR

These diodes, along with their associated resistors and capacitors, form a phase detector circuit. It might be recognised as an f.m. detector, but is more like the phase discriminator found in flywheel-controlled line timebases.

TR2 is a crystal-controlled oscillator, the reference subcarrier generator. The crystal is in the base circuit and it oscillates at a frequency of about 4·43 MHz when inductor L1 in the collector is properly tuned.

The sine wave signal at the collector is coupled to the base of the buffer amplifier, TR3. This is wired as a common-collector stage, and the reference signal is taken from its emitter and passed to the PAL detectors, thereby inserting the missing subcarrier for demodulation of the chroma signal.

A sample of this reference signal is fed back to the phase detector and applied via transformer T1. Should the phase, and hence frequency, of the reference

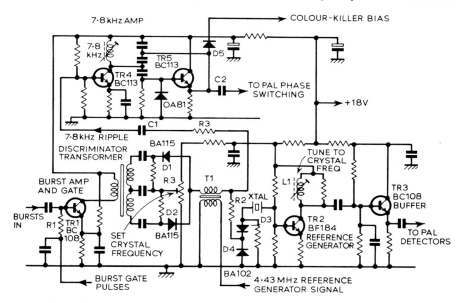

FIG. 9.19. *Burst, reference generator and indent amplifier circuits (Decca CTV25).*

signal tend to deviate from that of the colour bursts, the phase detector produces a d.c. output at the junction of D1 and D2. This is a control potential which pulls the phase of the oscillator signal until it matches that of the burst signal.

It does this by way of capacitor-diodes, D3 and D4, in the crystal circuit of TR2. These diodes decrease in capacitance as the reverse bias across them is increased. Now, while it is not possible to swing the frequency of a crystal oscillator very much, radio amateurs many years ago found that a small change in frequency could be achieved by varying the capacitance across the crystal at the expence of a certain amount of crystal 'activity'.

The diodes give this capacitance and the control potential from the phase detector provides the necessary change to phase-lock the oscillator. The control potential is applied to the junction of D3/D4 through the primary of T1 and R2. Preset R3 allows the standing bias on the diodes to be initially adjusted for the correct oscillator frequency. This is rather like the 'line lock' control associated with line flywheel circuits.

It will be recalled that at the transmitter the colour bursts are alternated in phase in synchronism with the line frequency to simulate the necessary PAL alternations. These phase alternations of the bursts at the phase detector produce a ripple signal at half line frequency (7·8 kHz) which has already been mentioned many times.

This is particularly strong at T1 primary and it is fed from this source to the base of TR4, via R3 and C1. The amplifier composed of TR4 and TR5 is, in fact, specifically designed to amplify this 7·8 kHz ripple signal. TR5 stage is another common-collector amplifier, the ripple being delivered to the PAL phase switching via the emitter and C2.

The ripple signal is also fed to D5, which rectifies it and turns it into a d.c. potential for biasing-on the colour-killer stage in the chroma channel, as we have already seen.

Buffers TR3 and TR5 are necessary to prevent loading affecting the reference generator and 7·8 kHz amplifier respectively. They also give low-impedance signal outputs, ideal for driving the chroma detectors and the PAL switching.

CHAPTER 10

Encoding and Decoding

ENCODING IS the process at the transmitter which puts the chroma information on to the luminance signals, thereby creating the composite colour signal. Decoding is the opposite action at the receiver, whereby the colour information is removed from the luminance information and separately 'detected'. This produces the original R − Y and B − Y signals for application to the colour-difference amplifiers. It must be understood that although the composite signal is demodulated by the vision detector, only the luminance signal at that point is at real video-frequency. The chroma information is still encoded and appears as subcarrier sideband components, which must be further demodulated (decoded) to yield the colour-difference signals.

The process of chroma demodulation is different from that in ordinary a.m. circuits because the video-frequency subcarrier is suppressed at the transmitter, leaving only the chroma sidebands—mixed in with the luminance signal. So a vital part of the demodulation process is the replacement of the missing sub-carrier.

The subcarrier, approximately 4·43 MHz, is generated at the receiver by the reference oscillator and is controlled in exact frequency and phase relative to the suppressed subcarrier at the transmitter by means of 'bursts' of true sub-carrier frequency generated on the back porches of the line sync pulses. These bursts are sometimes called 'colour sync'.

To understand chroma demodulation it is desirable to begin at the transmitter so that the nature of the chroma signal is appreciated. The method is rather special in that it has to put two signals together and then add them to the lumi-nance so that at any instant there is only one signal voltage—yet allowing all three to be sorted out at the receiver.

QUADRATURE MODULATION

The mixing and separating of the R − Y and B − Y signals is done by using quadrature modulation. Two subcarriers at 4·43 MHz are used, both derived from a common generator, but one lags the other in phase by 90 degrees. The two colour-difference signals are amplitude modulated on to these, producing two sidebands for each signal. The subcarriers are suppressed and the remaining sidebands added together giving one resultant pair of sidebands. These, like any such sidebands, are represented by a single signal which is added to the luminance signal.

The practical receiver incorporates various features, due to the special PAL techniques, which will be explained later. For the moment let us stick to the basic matter of demodulation which is equally applicable to NTSC and PAL.

In the set is a decoder which, when fed with chroma signal and reference signal, delivers the two colour-difference signals. The chroma signal is composed of two sine waves displaced in phase by 90 degrees, as illustrated by the full-line waves A and B in Fig. 10.1. When these waves are added together we get a third wave, the actual signal, shown by broken line C.

In Fig. 10.1 the phase displacement can be expressed as wave B lagging wave A by 90 degrees. The resultant wave is obtained by adding instantaneous amplitude values of waves A and B at many points along the horizontal axis. Only

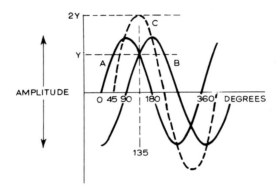

FIG. 10.1. *Waveform B lags waveform A by 90 degrees, and the* peak *of the resultant C is 1·4 times the* peak *of A or B.*

one such addition has been marked, at 135 degrees where waves A and B each has Y amplitude. Clearly, then, at the corresponding point along the horizontal axis wave C—the resultant—has 2Y amplitude*.

The diagram also shows the phase of the resultant wave relative to A and B, which is 45 degrees because waves A and B have the same amplitude and 90 degrees phase difference. A change in amplitude of either A or B wave (or both) changes the relative phase and amplitude of resultant C. This will become clearer later when we look at vector diagrams.

The decoder is essentially a chroma demodulator, as already mentioned, which implies that it contains some sort of 'detecting' or rectifying element. It does, in fact, contain two of these—one for each of the two colour-difference signals. Ordinary a.m. detection cannot be used, for this would only give a video

---

* *Peak* amplitude of C is 1·4 times that of A or B.

signal corresponding to the sum of the two colour-difference signals. A special technique is adopted.

The missing subcarriers have first to be replaced and, to match up with the conditions of quadrature modulation at the transmitter, the subcarrier fed to one detector must be shifted 90 degrees in phase relative to that fed to the other.

This is a simple matter, achieved by the use of transformer windings and R and C elements. We can now show the decoder (Fig. 10.2) as two separate detectors each receiving chroma signal and reference signal—which serves to replace the subcarrier—with 90 degrees phase displacement.

DETECTOR ACTION

The detector action is basically fairly straightforward, for the reference signal can be looked upon as switching the detectors on and off at the reference frequency (4·43 MHz). In the basic mode of operation, the R − Y detector is switched on just when the B − Y component of the signal is zero. So the signal it samples is R − Y alone. Similarly, the B − Y detector samples the signal only when the R − Y component is at zero.

In practice, distortion can result in the signal components *not* being at zero when the detectors operate. The sine wave in Fig. 10.3(*a*) represents the chroma signal at one instant in time, and this is sampled by the two detectors at different periods in time corresponding to the 90 degrees phase difference. When the red detector switches on, the wave amplitude is sampled at A, which is positive;

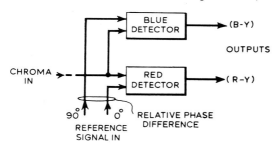

FIG. 10.2. *The blue and red detectors in the decoder are each fed with reference signal plus chroma signal.*

this switches off and 90 degrees later blue switches on and samples the amplitude at B, which is negative. The red detector thus gives a +(R − Y) output corresponding to amplitude A, and the blue detector gives a −(B − Y) output corresponding to amplitude B.

Figure 10.3(*b*) shows what may happen at another time. Here the red amplitude is −A and the blue amplitude −B. Each detector, therefore, can deliver positive or negative signals depending on the phase of the chroma signal. When the phase of the chroma signal changes relative to the reference signal, the amplitudes and polarities of the detector outputs also change.

Clearly, when the *amplitude* of the chroma signal changes, the amplitudes of *both* detector signals change together. This is brought out in Fig. 10.3(*c*), which shows the phase condition as in Fig. 10.3(*b*), but at reduced amplitude.

Previous chapters have revealed that the instantaneous hue of a scanned element of colour scene controls the phase of the chroma signal, while the saturation of the hue controls the amplitude of signal. It can now be seen how this is reversed at the receiver—the *phase* changing the relative red and blue signal amplitudes and the *amplitude* changing the signal amplitudes together.

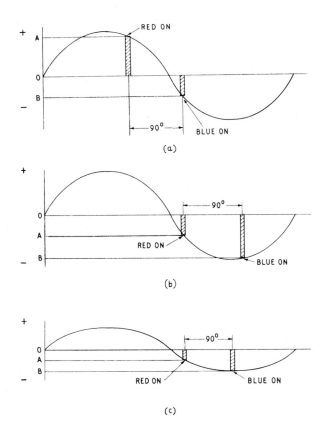

FIG. 10.3. *Basic action of the colour detectors. They are switched by and conduct on the peak value of the reference signal, and because of the 90 degree phase difference, one detector switches on while the other switches off. Chroma signal amplitude is thus 'sampled' first by one detector and then by the other at a rate corresponding to the frequency of the phase-locked reference signal. Curve (a) shows positive and negative 'samplings', curve (b) two negative 'samplings' and curve (c) the same as (b) but at reduced chroma amplitude.*

Zero chroma signal means zero colour-difference signals and zero colour on the picture tube—all three beams then being under the control of the Y signal alone.

This explanation, it must be noted, applies to NTSC and simple PAL. We could go straight on to the special PAL features but it is better to dwell for a few paragraphs on the reason behind them and to introduce vector treatment of the subject.

VECTORS

We have seen that phase is simply the term used to express the difference in time of the signal with respect to another or with respect to its own 'correct' time. With a succession of identical waves the only period of interest, from the point of phase, is one cycle. Thus, as we have seen, phase difference need not be expressed in time but as an angle within a circle (360 degrees). A signal which is late by a quarter of a cycle is said to be lagging by 90 degrees.

In a complete and far-spread system, like colour broadcasting and receiving, delays of signal occur all too easily, both in the circuits and in the transmission path. If the signal was uniformly delayed there would be no distressing consequences. Unfortunately, phase delay tends to occur at different frequencies and to varying extent.

The reason that phase delay matters more with colour TV than monochrome is because quadrature modulation makes phase and colour 'the same thing'.

To show this we must use the vector diagram. This is just a simpler way of showing Fig. 10.1, representing the two colour-difference signals and their sum. Instead of the entire waves, only one set of instantaneous values is represented.

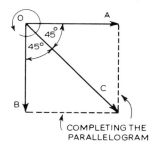

FIG. 10.4. *Vector representation of the waveforms in Fig. 10.1.*

The vector equivalent of Fig. 10.1 is given in Fig. 10.4, where lines *A* and *B* represent instantaneous values of signal with an angle 90 degrees between them.

The addition of *A* and *B* is now a simple matter of completing the parallelogram (dotted lines) and drawing the diagonal line *C* to represent the amplitude of the sum signal (when measured in the units used in drawing *A* and *B*) and also the phase angle with respect to both *A* and *B*.

The 'sense' of the diagram is said to be anticlockwise, indicated by the circled arrow at the corner, which is a way of showing that wave *B lags* wave *A* (by 90 degrees). In colour TV the resultant (*C* in Fig. 10.4) is called the *phasor*.

To see what happens during a complete cycle we would need a whole series of such diagrams showing the state of affairs instant by instant. It is much easier, though, to use our imagination and visualise the lines *A* and *B* varying in length and the phasor *C* not only changing in length but waggling to and fro.

The phasor in Fig. 10.4 can take up any position within the quarter of a cycle (90 degrees). Zero *B* signal would mean that the phasor was represented by signal *A* alone. Conversely, the phasor would be represented by signal *B* alone with signal *A* is zero. Fig. 10.5 shows (*a*) the position of the phasor when

A is small compared with *B*, and (*b*) when *B* is small compared with *A*. In (*a*) the phasor is only a small angle from *B*, and in (*b*) the phasor is only a small angle from *A*.

Signal polarity can be reversed easily as we well know. One way of achieving this is by changing round the connections to a coil—there are other ways, too.

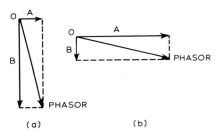

Fig. 10.5. *When one vector component is small and the other large the phasor is only a few degrees from the larger.*

(a)   (b)

Vectors, therefore, must be able to display 'reversed' signals. When one gives this a little thought it is obviously just a matter of turning one or both vectors through 180 degrees.

The four diagrams in Fig. 10.6 illustrate this. That at (*a*) repeats Fig. 10.4, (*b*) shows signal *A* reversed, (*c*) shows *B* reversed, and (*d*) shows *A* and *B* both

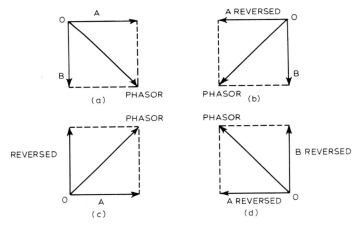

Fig. 10.6. *The phasor can be made to traverse 360 degrees by phase-changing each vector component separately or both together.*

reversed. The variations make it possible for the phasor to swing round a complete circle (360 degrees) according to the lengths and polarities of the vectors *A* and *B*. Fig. 10.7 combines the diagrams of Fig. 10.6 into a more rational one, and is set up to depict the condition of Fig. 10.4 and Fig. 10.6(*a*). Any of the other conditions can just as easily be presented.

Now let us leave this for a minute and get back to the transmitter, for we must tie this vector business up with the chroma signal. As will have become apparent, the phasor is not merely a concept but the representation of the actual chroma signal. The modulators, shown in Fig. 10.8, are designed so that when the colour information is zero the output is also zero. The colour-difference signals, anyway, resolve to zero when the scene lacks colour. They control the modulators so that output rises only when colour-difference signals appear.

BASIC MODULATOR

The basic circuit of one modulator, say, the R — Y, is given in Fig. 10.9. Another just like it exists for the B — Y signal. V2 and V3 are driven from the phase-splitter V1. When there is no R — Y input V2 and V3 conduct equally.

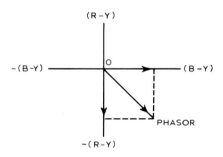

FIG. 10.7. *Fig. 10.6 elements arranged into one diagram, with the axes labelled to correspond to the colour-difference signals.*

Subcarrier is applied, via T1, to the third grids and, because T1 secondary is centre-tapped, the signal on one grid is 180 degrees displaced from that on the other grid. Subcarrier signals in the common load R1 thus add in anti-phase, giving zero output.

Presence of R — Y signal in V1 phase-splitter causes V2 and V3 conduction to unbalance—since the control grid of one goes towards positive while that

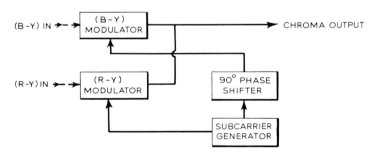

FIG. 10.8. *How the chroma signal is produced at the transmitter.*

of the other goes negative—and signal appears across R1. This represents the sum of the upper and lower sidebands of the R — Y chroma signal. Clearly, the amplitude of the modulated signal varies in sympathy with that of the R — Y input.

Outputs from *two* such modulators are combined at 90 degrees—owing to the phase-shift in the subcarrier signal to one of them (see Fig. 10.8). Because of this we can in Fig. 10.7 label one axis R − Y and the other B − Y. The sum, represented by the phasor, embodies all the chroma information. Note that each axis has plus and minus directions from zero at the centre.

When there is no colour information the vector lines both go to zero and the phasor likewise shrinks into the zero point in the middle of the diagram; when there is only +(R − Y) information it goes up the vertical axis, and when there is half maximum +(B − Y) and half maximum −(R − Y) it assumes the position drawn on the diagram. *Maximum* B − Y *and* R − Y *on this*

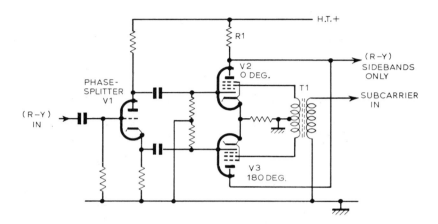

FIG. 10.9. *Section of basic quadrature modulator. Output increases with increase in amplitude of the colour-difference signal, and output is zero when input is zero. Each colour-difference signal has such a modulator, the outputs of which are combined.*

diagram is only arbitrary, of course, but it has been indicated by the length of the R − Y and B − Y axes along which the corresponding vectors operate.

In the course of normal working the phasor can be imagined to jump around the 360 degree path, giving the *hue* of each picture element in turn. The length of the phasor at each position is a measure of the *saturation* of the colour at any instant.

It is possible to show mathematically in a somewhat abstract manner that the resultant of the sidebands (due to amplitude modulation of the two colour-difference signals on the subcarrier) adds vectorially to the carrier at every instant to produce the modulated wave. It could also be shown that at all times the resultant sideband signal, formed by the addition of the upper and lower sideband vectors, is either in phase or 180 degrees out of phase with the subcarrier.

This means that we can treat the resultant sideband signal by itself as a signal that may be in phase or 180 degrees out of phase with the carrier. We transmit only the sideband signal, and the very important point here is that the 'phase reference' is obtained by replacing the subcarrier at the receiver.

There is one more aspect of all this: although the subcarrier is suppressed, we still talk about subcarrier on the chroma signal! Clearly, of course, the chroma signal has the subcarrier frequency—but it is not really the subcarrier, for we know that this has been suppressed. We say we get 'subcarrier beat' troubles when the picture is highly colour saturated. This is when the output from the modulators is high (a lot of $R - Y$ and $B - Y$ input) and the phase and amplitude change (due to the action of the sidebands), but not the frequency.

COLOUR RELATIONSHIP

One feature of the colour-difference system is that the vector diagram of Fig. 10.7 can be placed upon a 'standard' chromaticity diagram—suitably modified for the tube colour phosphors—in a particular orientation and with vector corresponding to the white region in the middle, and the tip of the phasor can then be visualised as actually pointing to the colour. This is shown for one instant in Fig. 10.10, where the phasor is indicating fairly highly saturated green, which is given by the $-(R - Y)$ signal being almost full amplitude (relatively) and the $-(B - Y)$ signal being about half amplitude.

Pink would be given by about half the amplitude of $R - Y$ signal only (but possibly with a trace of $-(B - Y)$ signal) while red would be given by a

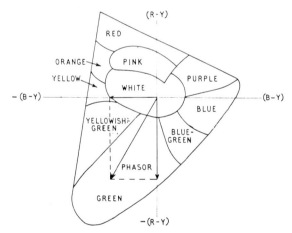

FIG. 10.10. *How the colour-difference vector relates to the colour diagram. The phasor is indicating green at about half saturation.*

phasor of the same angle but of greater length. In other words, pink is de-saturated red. All pale colours are given by short phasors while zero length is white.

We can see how any phase distortion results in colour error. If, for example, the phasor was put in error by about 30 degrees, due to external or internal phase distortion, the display at that instant would change from correct green, as indicated on the diagram, to a yellow. This brings us to the PAL action.

Let us suppose that on one line of picture and at the instant represented by the full-line phasor in Fig. 10.11(a), a yellow picture element is scanned, but that the system phase-error is causing the phasor to take up the broken-line position corresponding to a red display.

With PAL, the phase of the R — Y signal is reversed on the next line as shown in Fig. 10.11(b). The error angle is unchanged but the phase reversal changes the colour from what would have been yellow to towards green.

At (a) the 30 degree phase error changes the display to 140 degrees (170 — 30), while the R — Y phase reversal at (b) changes the display to 200 degrees (170 + 30). The average of the two, therefore, is 170 degrees, as shown at (b), which is the angle for the correct yellow display.

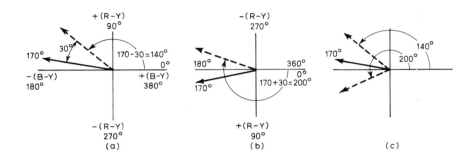

FIG. 10.11. *A yellow element at 170 degrees on one line is displayed as towards red due to a 30 degree phase error (a) and on the next as towards green due to the complementary error resulting from the (R — Y) phase reversal on that line (b). When the error and complementary error displays or signals are averaged by the eye or electronically the correct colour at 170 degrees is yielded (c).*

Hue correction in PAL-S, therefore, is a function of the eyes of the viewer, the eyes mixing the two displays in terms of lights, as with the glowing dots on the shadowmask screen, so as to give the impression of true yellow. This is shown in Fig. 10.12. With PAL-D, the two signals are averaged electronically by the delay-line decoder so that the signal corresponding to the correct colour is produced, in spite of phase error, as explained later.

Fig. 10.13 shows the PAL switching at the transmitter. This is responsible for the PAL line R — Y phase reversals. An inverter is interposed in the 90 degrees phase-shifted subcarrier signal path to the R — Y modulator, and this is switched by pulses from the line circuits. The bursts, too, are alternated ±45 degrees about the — (B — Y) axis (see Fig. 10.20). This is why they are called 'swinging bursts'.

HANOVER BLIND EFFECT

The subjective correction given by PAL-S produces picture interference when the phase errors are significant. This effect is called a 'Hanover blind' because it was first seen in the Telefunken laboratories where the PAL system was developed, and its cause is the interlacing of alternate field scans. The interlaced lines carry opposing colour errors, which means that two adjacent lines carry the same error colour, while the next adjacent lines carry the opposing colour error.

When the phase error is fairly large, a substantial difference in colour exists two lines thick—and the eyes are unable to integrate two pairs of lines in terms of the true colour. The subjective shortcomings are eliminated by PAL-D, which uses a signal delay line arranged in a manner that permits electrical cancellation of phase error.

A PAL-D decoder block diagram is given in Fig. 10.14, and it is instructive to compare this with the basic non-PAL decoder in Fig. 10.2, which uses just a

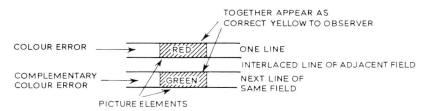

Fig. 10.12. *The eye integrates the real error (red) and its complementary (green) and discerns the true yellow as transmitted (see Fig. 10.11). Interference, called Hanover blinds, is produced in this system when the phase error is large.*

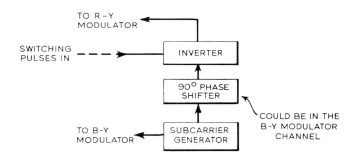

Fig. 10.13. *PAL switching at the transmitter.*

pair of colour detectors and a reference signal. Fig. 10.14 still has the two colour detectors, the reference signal input to them, to one via the 90 degree shifter, and the colour-difference signals at their outputs. There are several extra items, as can be seen.

A very important one is the PAL delay line—a highly engineered device working on the ultrasonic principle. It uses a specially processed block of glass through which the lines of chroma signal pass after being translated to ultrasonic vibrations.

There are two transducers, one acting in a similar manner to that of an ultrasonic 'loudspeaker' and the other to that of an ultrasonic 'microphone'—input and output transducers, respectively. The input transducer receives the chroma signal, changes it to a supersonic wave and sends it through the glass block. The pressure wave operates the output transducer which converts it back to the electrical chroma signal. The purpose of this is to delay the chroma signal by the time period of one line -63.943 ± 0.003 μs, loosely referred to as 64 μs.

It is less complicated to secure this magnitude of delay by ultrasonic than by electrical methods. Fig. 10.15 illustrates the principle behind the delay line, used in many British sets, which itself is illustrated in Fig. 10.16.

The delay line feeds the chroma signal into the matrix (Fig. 10.14) and the chroma signal is also applied direct. The matrix is nothing more than a circuit network which processes the two signals by adding them together and by subtracting one from the other.

This matrix must not be confused with the matrix in the colour-difference circuits that derives the G — Y signal. That one has nothing to do with that shown in Fig. 10.14, which is sometimes called by other names—'mixer', 'adder/subtractor' and so forth.

Because the chroma signal fed into the delay line appears at the output exactly one line-period later, the matrix receives the signal of, say, line 3

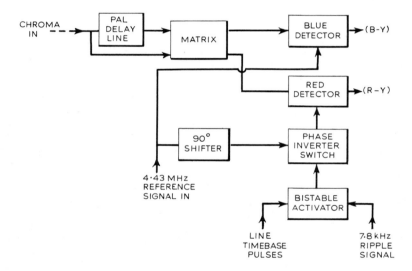

FIG. 10.14. *Basic elements of the PAL decoder.*

(direct) together with the signal of line 1 (delayed) in exact synchronism, as governed by the accuracy of the delay line. Remember that the fields are interlaced so that the line *numbers* are not consecutive although the signals are. Similarly, the matrix receives the signal of line 3 (delayed) with line 5 (direct) and so on.

We must not forget that the polarity scale of the R — Y chroma signal is changing line by line at the same time. And, with the matrix both adding and subtracting, we get a very useful effect—the separation of the R — Y and B — Y components *before* detection. These signals differ, moreover, from the original variable-phase chroma signal in that the phase of each one is fixed.

Vectors show why this is so. Fig. 10.17 shows at (*a*) the signals on, say, line 5, and at (*b*) on line 7 (the next consecutive line in the field). When (*a*) is subtracted from (*b*) we get configuration (*c*), where the two B — Y components cancel out, leaving only the $\pm$R — Y components, which add to each other, giving ± 2(R — Y).

When (*a*) is added to (*b*) we get resultant (*c*), where the $\pm$R — Y components cancel out and the B — Y components add, giving 2(B — Y). The net result

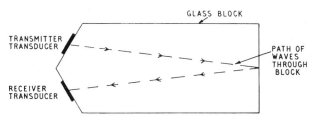

FIG. 10.15. *PAL delay line works on the ultrasonic principle. Blocks are made slightly longer than required to allow the reflecting end to be ground down to achieve the exact delay (Mullard).*

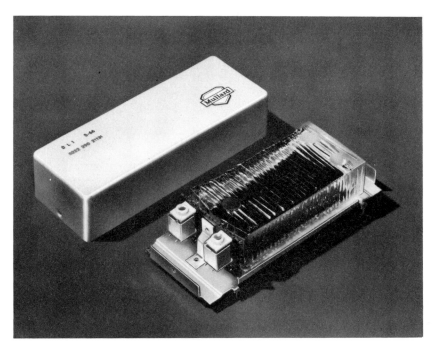

FIG. 10.16. *PAL delay line with termination presets (courtesy, Mullard Ltd.).*

is that phase errors are averaged over two lines, so that the two detectors receive a mean phase-correct input.

The signals represented in Fig. 10.17, however, are 'phase-perfect' and fail to take into account the reference signal, which must, of course, be applied to the detectors for them to work.

Phase error can be looked upon as meaning that the chroma signals fail to line-up exactly with the R − Y and B − Y axes. However, although the signals may be 'wrong' when sampled by the synchronous detectors, as each detector is operating only on a signal of its own colour, no mixing of the other colour occurs, and so there is no shift of hue.

Fig. 10.18 represents how the system phase error is cancelled out as a by-product in reduction of output from R to r and B to b. This is far better than the phase error which changes hue, for it only reduces the saturation a little (because both red and blue outputs fall equally together).

It is imperative in the PAL decoder (Fig. 10.14) that the red detector is switched in some way so that it detects +(R − Y) on one line and −(R − Y) on the next. Sad things would happen to the picture if the detector 'detected', say, a +(R − Y) line for a −(R − Y) line.

The detector thus has to be 'phase-switched', as at the transmitter modulator, and this is accomplished simply by phase-alternating the reference signal applied to it. Fig. 10.14 shows that the reference signal to the red detector is

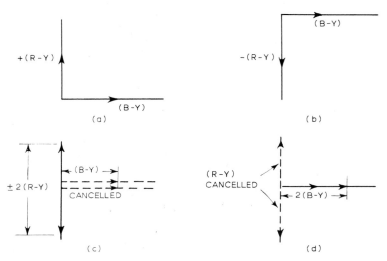

FIG. 10.17. *Vectors showing the operation of the PAL delay line. (a) and (b) represent alternate lines, (c) gives the result of subtraction and (d) the result of addition, yielding ±2(R − Y) and 2(B − Y) respectively.*

passed through a 'phase inverter' switch, which alternates the phase of the reference signal in synchronism with the R − Y alternations applied at the transmitter.

The phase alternator or inverter switch is activated by a bistable circuit, rather like a multivibrator with the oscillatory sustaining feedback removed. It fails to oscillate (switch) by itself and needs pulses to alter its conduction state. These pulses are supplied by the line timebase and, to ensure that the bistable 'latches on' to the correct R − Y line (plus or minus), it is 'synchronised' by the 7·8 kHz ripple signal appearing in the phase detector, as the result of ±45 degrees phase swings applied to the colour bursts.

Let us now turn to the main circuits in the decoder, which include the delay line and its amplifiers, PAL switching devices, and the red and blue detectors. The other parts of the decoder are fully examined in the previous chapters.

DECODER CIRCUITS

Fig. 10.19 shows the circuit of a decoder which forms the basis of decoders used in some commercial sets. It is best to consider the circuit in two parts, the top part dealing with the PAL switching and the bottom with the delay line and the associated matrix. Let us start with the bottom part first. The stage

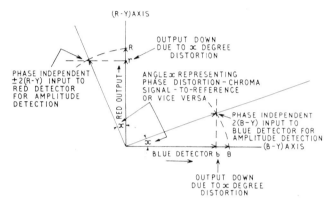

FIG. 10.18. *Vectors showing how PAL-D cancels phase error. Unwanted changes in phase produce detector* amplitude *differences, giving slight reduction in saturation instead of variation in hue.*

round TR1 is an amplifier which receives the filtered chroma signal at its base. It is really an extension of the chroma amplifier channel with an interposed 6 MHz filter for tuning-out any intercarrier sound signal, and is referred to as the 'chroma driver' or 'delay-line driver'.

TR1 should be considered in conjunction with the delay line, which represents the collector load. Thus, the collector signal is delayed chroma appearing at the output of the delay line. The emitter also delivers the chroma signal direct—without delay. This is summed and differenced with the delayed signal by the alternating phase $R - Y$ signal component in the matrix action of the circuit.

Transformer T1 comes into the adding and subtracting business. This is bifilar-wound and delivers delayed chroma signals of opposite polarities at points A and B. These are added to and subtracted from the direct signal at the junction of R1 and R2 by the alternating $R - Y$ phase, as already explained, resulting in corresponding chroma signals at the outputs A and B for application to the appropriate detectors. The red and blue signals are effectively amplitude modulated, the phase factor having been neutralised by the adding and subtracting process, as already explained.

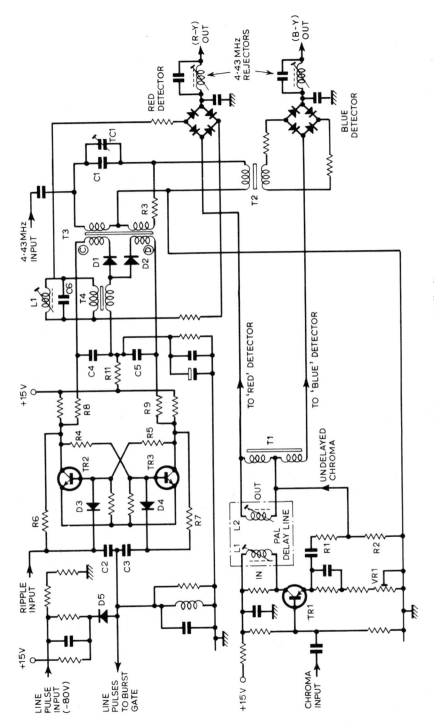

Fig. 10.19. *Decoding circuit (Mullard).*

Inductors L1 and L2 relate to the delay line, and they are actually incorporated in the component itself for tuning the input and output impedances for the correct delay and bandwidth characteristics relative to a 150Ω termination impedance. They are not really adjustments for the service technician, but more in the nature of 'factory presets'. Their incorrect adjustment, however, would change the load impedance, alter the bandwidth and hence change the delay characteristic, impairing the PAL action, and causing Hanover blind interference.

DECODER MATRIX ADJUSTMENT

The preset resistor VR1, in the emitter of TR1, serves to adjust the gain of the stage so that the amplitude of delayed signal equals that of the undelayed signal at the R1/R2 junction (unbalance results in Hanover blind interference). This sort of adjustment can be made by applying a signal exactly at the subcarrier frequency to the chroma input and then monitoring it on an oscilloscope connected to A or B on T1. The subcarrier frequency can be picked up from the subcarrier reference generator buffer in the decoder and fed to the chroma input through a 100 kΩ resistor. To ensure that the frequency is accurately phased, the set should be picking up a colour signal when this adjustment is performed.

With the oscilloscope's Y input connected to T1 (Y attenuator to about 100 mV/cm), adjustment of VR1 will produce either a signal peak or null, depending on whether the connection is to A or B. The side giving the null is the best for critical adjustment. The detector connections are first removed.

The top right-hand part of Fig. 10.19 shows the 4·43 MHz reference signal input to the red and blue detectors (the detectors themselves will be considered later). This reaches the blue detector via T2 and the red detector via the double transformer T3 and T4.

It will be recalled that the reference signal is extracted from the emitter circuit of the common-collector buffer stage following the reference generator, giving a low-impedance drive to the detectors.

The reference signal to the blue detector must be 90 degree phase-shifted relative to the reference signal to the red detector, and this is achieved by the action of R3, C1 and TC1 in conjunction with the secondary of T3. The reference signal to the red detector, on the other hand, must be phase alternated to match the PAL alternations. This action switches the red detector line by line in accordance with the polarity changes of the red chroma signal.

RED DETECTOR SWITCHING

Transistors TR2 and TR3 are arranged in the form of an electronic switch, which alternately turns diodes D1 and D2, connected to the primaries of T3, on and off. The two transistors are cross-coupled—base to collector in each case—by R4 and R5, an arrangement which has something in common with a multivibrator circuit, such as the kind found in timebase oscillators. The difference is that the coupling capacitors are missing in Fig. 10.19, and the circuit is slightly modified with diodes.

In this form it is called *bistable* circuit, which means that it fails to oscillate by itself, as does the common multivibrator. Nevertheless, the circuit can be made to switch so that TR2 is on when TR3 is off, and vice versa. Pulses are needed to obtain this action. Bistable means that the circuit is inherently stable in two modes—TR2 on and TR3 off *and* TR2 off and TR3 on.

There is another species of this kind of circuit, called *monostable*, which means that it automatically (due to feedback) switches to one mode, but needs a pulse to switch it to the other, after which it switches back to the first mode by itself.

Diodes D3 and D4 along with the associated components C2, C3, R6 and R7, change the basic bistable circuit into a binary counter of the kind used in computers, and the trigger input is applied to the junction of C2 and C3. Diodes D3 and D4 are controlled by the transistor collector voltages. When TR2 is on, it is bottomed, so that D3 is just conducting. TR3 is then off, resulting in D4 being biased-off by the positive collector voltage. The next switching pulse is passed to TR2 base, switching this transistor off and TR3 on.

Time-constants C2,R6 and C3,R7 are arranged to hold the diodes conducting until the finish of the input pulse, thereby ensuring that the circuit always completes its switching before the 'gate' opens to the next input pulse. It is worth remembering that a bistable circuit performs one complete cycle for every two input pulses; that is, one pulse appears at each collector during one complete cycle.

BISTABLE SWITCHING DIODES

Now let us see how the bistable switch actuates diodes D1 and D2 connected to transformer T3. A d.c. circuit from TR2 collector to D1 is provided through R8 and a circuit from TR3 collector to D2 through R9. The 'anode' of each diode is connected to a positive potential at the junction of the potential divider formed by R10 and R11. Whether the diodes are conducting or non-conducting, therefore, depends on the 'cathode' potential from the transistor collectors.

When TR2 is on, the potential from its collector to D1 'cathode' is less than that D1 anode and the diode conducts. At this instant, TR3 is off and its collector potential is maximum positive. Since this is communicated to D2 'cathode', this diode is biased off. The conditions reverse on the next switching pulse—D1 off and D2 on—and so on from pulse to pulse.

The 4·4 MHz reference signal arrives at the red detector through T3 and T4, but its phase at T4 primary depends on whether it is derived from winding C or winding D on T3. Clearly, when D1 is conducting it is derived from winding C via C4 and when D2 is conducting from winding D via C5.

One winding gives the reference signal in opposite phase to the other winding, but the signals are otherwise the same. Thus each time the bistable is switched by an input pulse the phase of the reference signal applied to the red detector is reversed, which is the requirement for PAL decoding.

The bistable is triggered by line pulses (−80 V from the line timebase) applied to the junction of C2 and C3, via the pulse-shaping circuit consisting of diode D5 and associated resistors and capacitor. This means that the phase of the red detector drive signal alternates line by line.

So far so good, but there is only a 50:50 chance that the switched phase will coincide with the line phase at the transmitter. If the switched phase of the red detector does not correspond to the phase of the original signal, the colours displayed are desaturated and tend to be the complementary of those transmitted. The picture could then appear rather like a poor negative of a colour photograph.

SWINGING BURST AND RIPPLE SIGNAL

The problem of phase identification is solved by using the 7·8 kHz ripple signal appearing in the phase detector circuit for synchronising the bistable to the correct phase. Some designs derive the ripple signal from the collector of a d.c. amplifier in the reference generator control circuit.

Before we can fully appreciate why the 7·8 kHz signal carries information on the phase of the transmitted red signal, we must once more recall the situation at the transmitter.

Here the colour bursts are swung in sympathy with the PAL line phase reversals ±45 degrees with respect to the −(B − Y) axis, hence the term *swinging bursts*. The sequence of events is revealed by the diagrams in Fig. 10.20, which show the position of the bursts during odd (non-phase-reversed) lines at (*a*) and during even (PAL phase-reversed) lines at (*b*).

HOW THE RIPPLE SIGNAL IS PRODUCED

The burst can be considered as a carrier which is phase-modulated by the 'swinging action'. The phase modulation is 'detected' by a phase detector and a 'ripple output' at half the line frequency (7·8 kHz) is produced. This is rather like an elongated square wave centred on a d.c. datum line, rising positively on the leading stroke and negatively on the falling one. Each half section of this wave is thus equal to half the line frequency, brought about because the zero reference phase of the burst is 'wobbled' ±45 degrees on alternate lines relative to a mean of −180 degrees from the 'blue phase', while the red signal is switched ±90 degrees relative to the 'blue phase'.

The ripple flicks a high-Q circuit into 'ringing' and, since this is tuned to half the line frequency, a sine wave of this frequency is produced, and it is this which operates the colour killer.

The ripple signal is also applied to the bistable switch that controls the red detector, and is colloquially referred to as the *colour ident signal*, meaning colour identification signal.

How does this 7·8 kHz ripple signal cause the bistable to trigger on the correct lines? The origin of the ripple signal is shown in Fig. 10.21. The transistor is arranged in the common-collector mode, and the ripple, which is fed from a high-Q tuned circuit to its base, appears at low impedance at the emitter.

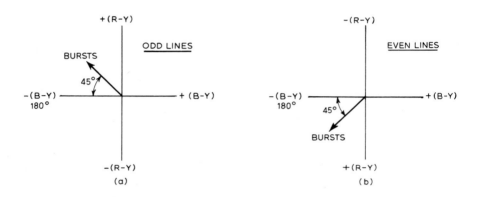

FIG. 10.20. *Showing how the bursts swing ±45 degrees relative to the (− B − Y) axis in sympathy with the PAL line phase reversals. This puts the average phase of the bursts along the − (B − Y) axis, which can be regarded as 180 degrees.*

The output of diode D1, in the emitter circuit, connects directly to the 7·8 kHz input on Fig. 10.19 (top left-hand corner), and the diode in conjunction with C1 and R1 (Fig. 10.21) provides phase-switch comparison which modifies the operation of the switching circuit, R6 and C2 in Fig. 10.19, as explained below.

When the ident signal indicates the correct phase the bistable line-pulse triggering is not affected, but if the phase happens to be incorrect, the line driving pulse is suppressed by a blocking bias produced by D1 in Fig. 10.21. The bistable then misses a count, thereby automatically correcting the phase.

It is interesting to note that this method of synchronising is so effective that it is normally better than the locking ability of the field timebase under very bad signal conditions. This results from the remarkably high signal-to-noise performance of the colour burst, the phase alternations of which, incidentally, reduce its effective amplitude to about two-thirds of that of the NTSC burst.

The peak value of the 7·8 kHz signal at the emitter of the transistor in Fig.

10.21 is rectified by a separate diode (not shown on the circuit) to produce the colour-killer bias. It will be remembered that this is used to turn on the chroma channel only when colour information (burst signal) is present on the transmission. At all other times the chroma channel is muted.

It is sometimes asked how the bursts can 'lock' the reference generator to a constant phase, corresponding to the suppressed subcarrier at the transmitter,

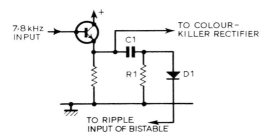

FIG. 10.21. *Source of 7·8 kHz ripple signal (Mullard). This circuit also feeds the colour killer rectifier.*

when they themselves are swinging ±45 degrees. The answer is that the phase detector has a relatively long time-constant, making it much too slow to follow the line-by-line swings. It thus establishes a mean control potential relative to the B — Y phase.

Returning now to the circuit in Fig. 10.19, the trimmer TC1 across T3 allows for accurate adjustment of the 90 degree phase shift between the R — Y and B — Y components, while L1 and C6 tune out harmonics from the switched red detector drive signal.

The red and blue detectors are called diode-bridge detectors or demodulators, because they each employ four diodes in the form of a bridge circuit. These work in a similar way to NTSC detectors, already explained, but with PAL the phase component of the chroma signals is cancelled out, as we have seen. This means that the detectors only have to 'detect' amplitude modulation, but this has to be detected in accurate time sequence as provided by the reference drive signals.

In other words, the reference signal switches the detectors on at exactly the right time for sampling the chroma information and for deriving R — Y and B — Y video components related to the amplitude of the signals at the time of sampling.

Any 4·43 MHz reference signal getting through the synchronous detectors is blocked by the rejector circuits (parallel tuned circuits) connected at their outputs.

BRC 2000 SERIES DECODER

The block diagram in Fig. 10.22 corresponds to the decoder of the all-transistor colour set in the BRC 2000 series. This differs a little from the circuit that we have been considering, notably in that a post delay line amplifier is employed. The circuit, too, differs in detail, as revealed in Fig. 10.23.

The chroma channel effectively terminates at the delay-line driver TR1, which feeds into the delay line at its emitter through C1 and to the detectors directly from its collector through C2. The direct chroma signal is applied to the red detector through R1 and to the blue detector through R2, while the

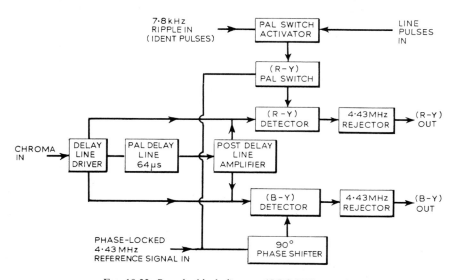

FIG. 10.22. *Decoder block diagram (BRC 2000 series).*

delayed signal is fed to the detectors respectively through R4 and R3 from the emitter of TR2 (common-collector stage), via C5. C4 puts the collector of this stage at earth potential with respect to the signal.

Feeds A and B to the red and blue detectors, therefore, receive the sum and difference signals, with R1, R2, R3 and R4 acting as a matrix (adder/subtractor). The signals can be balanced by the matrix preset, which simply adjusts the strength of the direct signals relative to the delayed ones. This is a factory preset adjustment.

The reference signal drive (4·43 MHz) is fed to the blue detector through a phase-shift network comprising L1, C8, T2 and the parallel resistor. The red detector drive is phase alternated by means of diodes D1 and D2 being switched on and off alternately by the PAL switch activator, which is a bistable circuit. Switching is by TR3 and TR4, via R5 and R6, but in this circuit T1 primary is a single centre-tapped winding.

With D1 'on' the reference signal passes through section *x* of T1 primary via C6, and with D2 conducting the signal passes through section *y* via C7. The red drive thus alternates in phase in the secondary as shown by (i) and (ii).

For ident, the 7·8 kHz pulses are first squared by TR5 and then fed to the base of TR6, which also receives line switching pulses. The collector output of TR6, which is a pulse mixer, switches the bistable. If the signal phase is in-

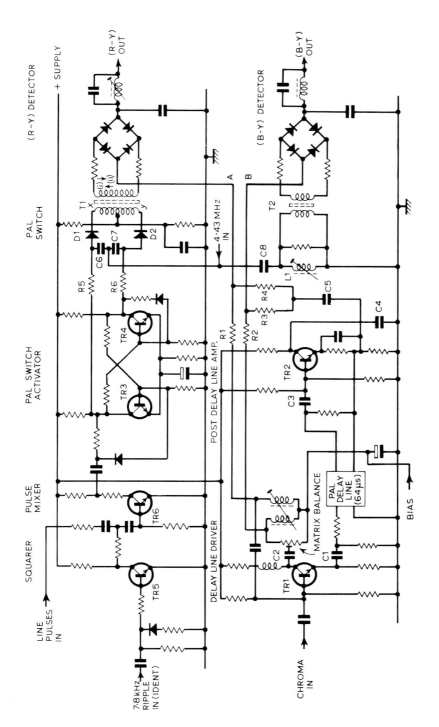

FIG. 10.23. *Transistorised decoder (BRC 2000 series).*

correct, the 7·8 kHz pulse cancels out the line triggering pulse in the mixer and there is no switching output. The bistable thus misses a count and then comes into phase. 4·43 MHz rejectors are used at the detector outputs as in the Mullard circuit. This, in fact, is a common feature of all synchronous detector circuits.

U AND V SIGNALS

All through this book we have looked upon the R − Y and B − Y signals as those that actually modulate the transmitter subcarrier. In practice, however, these signals are modified before they are applied to the modulators and so far as the PAL system is concerned they are called respectively the V and U signals.

The essential difference is that of amplitude—the relative strengths of both the V and U signals being less than the R − Y and B − Y signals. This is to avoid excessive modulation depth when the composite signal (luminance plus chrominance) is put on the r.f. carrier.

Fig. 10.24(a) shows the nature of the combined chrominance signal—the modulation envelope—due to a test colour bar pattern as delivered by the

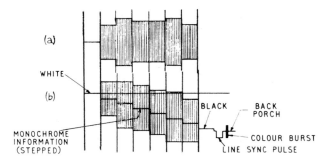

FIG. 10.24. *Chroma signal produced by the camera scanning 'standard' colour bars (a). Waveform (b) results when the chroma is added to the Y signal resulting from the same camera input.*

4·43 MHz subcarrier to the transmitter. Fig. 10.24(b) shows a likely addition of (a) to the luminance (Y) signal—note how the colour bars give a stepped grey-scale.

Fig. 10.25 shows with some exaggeration the effect when the colour information, superimposed on the monochrome information, causes the composite amplitude to rise above that required for full modulation of the carrier. Should this happen, of course, the transmitter would overload and the signal would suffer distortion. The problem is solved in practice by the application of 'weighting factors' (attenuation) to the B − Y and R − Y signals before they are applied to the colour modulators and this is done by the transmitter matrix.

Two factors are taken into account by the weighting: one is signal-to-noise performance and the other transmitter overload. If the signals are attenuated too much, colour noise becomes troublesome. However, on a normal picture the possible modulation overload is not so severe as indicated by fully saturated colour bars and so the weighting does not have to reduce the amplitude excessively.

Great consideration was given to this problem during the development of the PAL system and the weighting standard now adopted in Europe and on our own PAL system is expressed as:

$$\frac{B - Y}{2 \cdot 03} = U \quad \text{and} \quad \frac{R - Y}{1 \cdot 14} = V$$

Thus the B − Y signal is renamed the U signal and the R − Y signal becomes the V signal.

The signals can also be expressed as U = 0·493(B − Y) and V = 0·877(R − Y). This amounts to the same thing, which is that the B − Y signal is approximately halved in amplitude and the R − Y signal is reduced by a little over 10 per cent.

At the receiver the correct signal balance must be restored. This requires the U and V signals delivered by the synchronous detectors to be converted back

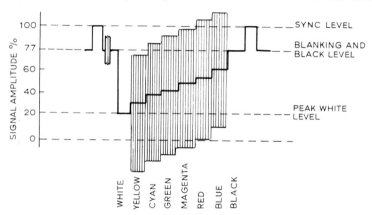

FIG. 10.25. *Exaggeration of 'over-modulation' due to the addition of unweighted chroma to the Y signal.*

to the B − Y and R − Y signals prior to application to the colour tube. This is achieved by tailoring the effective gains of the colour-difference channels to multiply the U signal by 2·03 and the V signal by 1·14.

Fig. 10.26 shows the composite PAL waveform radiated by the BBC during the transmission of the colour bars. It will be seen that the overall amplitude is within the 100 per cent modulation limits for 95 per cent saturated primary and complementary colours.

In explaining earlier how R − Y and B − Y signals add vectorially to give a chrominance signal (phasor) of varying amplitude and phase angle, it was implied that the colour-difference signals are at their 'natural' values. In fact, however, since they are weighted, the phasor is altered—not only in amplitude but also in phase.

At this juncture, we need reminding that the effective gains of the colour-difference channels are also adjusted to compensate for factors other than the U and V weighting. The signal specification for fully saturated colour bar signals in terms of *colour-difference* are red = 0·7Y, blue = 0·89Y and green = 0·41Y.

For correct grey-scale with cathode drive, taking into account picture tube spreads, the full drive voltages are in the order of red cathode 100 V, blue

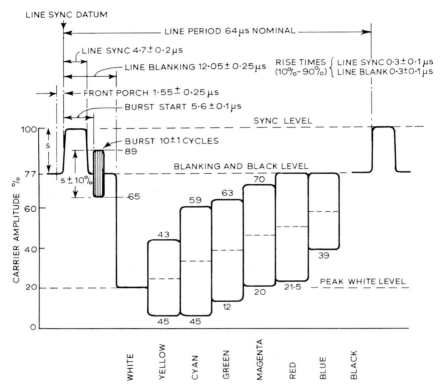

Fig. 10.26. *Colour waveform showing line sync pulses, colour burst and colour bar signals for full-amplitude 95 per cent saturated primary and complementary colours (BBC).*

cathode 94 V and green cathode 96 V, and translating into peak-to-peak colour-difference drives we get

$$R - Y = 2 \times 0{\cdot}7 \times 100 = 140 \text{ V},$$

$$B - Y = 2 \times 0{\cdot}89 \times 94 = 167 \text{ V},$$

$$G - Y = 2 \times 0{\cdot}41 \times 96 = 79 \text{ V}.$$

In practice, a further allowance of about 30 per cent must be made for the lower sensitivity of grid drive compared with cathode drive, which gives red, blue and green colour-difference peak-to-peak drives respectively of 183 V, 218 V and 102 V.

Taking into account the U and V signal weighting, which can be looked upon as the blue colour-difference signal having an amplitude of about 56 per cent of that of the red one, it is clear that the gain of the B − Y amplifier must be about twice that of the R − Y amplifier. However, to balance out spreads in the tube and components, most colour-difference amplifiers feature preset level controls, as we have seen.

Of course, since the green colour-difference is derived from a matrix between the red and blue colour-difference amplifiers, the correct green drive is likewise obtained from a green level preset.

I AND Q SIGNALS

It is noteworthy that the NTSC system employs I and Q signals as distinct from the U and V signals of the PAL system. These signals, too, are specified in terms of proportions and signs of R − Y and B − Y but the weighting factors are different.

The modulation axis of the I and Q signals also differs from that of the U and V signals. Any modulation axis can be adopted without altering the basic system provided, and the receiver's reference signal is phased accordingly. The reason for the chosen NTSC I axis is that it lines up with the orange and cyan zones on the colour diagram, which correspond approximately to the colours in which the eye is most sensitive to detail. The CCIR standard of the NTSC system puts the I signal into a 1·6 MHz bandwidth and the Q signal, which is the least colour sensitive, into the reduced bandwidth of 800 kHz.

The PAL system originally utilised the I and Q axes of NTSC but they were later shifted to simplify the receivers and both the U and V signals are given a bandwidth of 1 MHz, transmitted at not more than 3 dB down at 1·3 MHz and at least 20 dB down at 4 MHz.

It will be recalled that the R − Y axis (PAL V axis) lies across the red and green zones of the colour diagram, while the B − Y axis (PAL U axis) lies across the yellow and blue zones.

The PAL U and V axes do not change the manner in which the phasor operates which is like a 'pointer' moving round the colour diagram with changes in U and V amplitude and polarity. However, its amplitude is now expressed as:

$$\text{Phasor amplitude} = \sqrt{(U^2 + V^2)}$$

For example, on saturated green of the colour bars U = 0·29 unit and V = 0·52 unit, thereby giving the phasor an amplitude of 0·59 unit.

Remember, too, that it is the V signal which is reversed in sign line-by-line in accordance with the PAL alternations.

The waveforms in Fig. 10.27, showing the relationships when fully saturated colour bars are scanned, are useful to analyse the system as a whole. It should be remembered, though, that for most of the time *picture* colours are not fully saturated. While the 'primaries' always contribute to the Y signal in the proportions 0·3R, 0·59G and 0·11B, they mostly have three different values below the 'unity' value represented in the diagram.

At the top are indicated the colour bars plus white and black, and below are shown the resultant signals from the blue, red and green tubes in the pick-up camera. At (*e*) is shown the luminance signal derived by taking these primary signals in the proportions of 0·3R, 0·59G and 0·11B.

The next step is the development of the colour-difference signals. These are the ones which are quadrature-modulated on to the 4·43 MHz subcarrier.

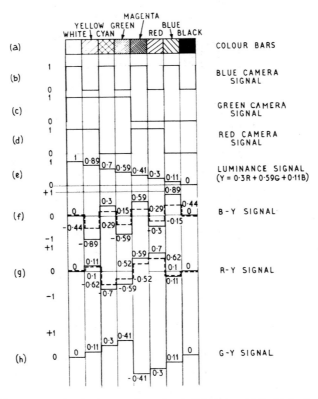

Fig. 10.27. *Blue, green and red primary-colour signals (b), (c) and (d) obtained when the camera scans a 'standard' colour bar pattern, such as (a). Luminance signal (e) is derived from correct proportions of each primary-colour signal. Blue and red colour-difference signals are shown at (f) and (g), and the dotted lines in these correspond to the weighted values, giving the U and V signals respectively. The green colour-difference signal is shown at (h). These are very important colour TV signals. The relative signal levels refer to full modulation and 100 per cent saturation. Different levels arise from different modulation level and saturation.*

Obviously, for the B − Y colour-difference signal, waveform (*e*), the Y signal, has to be subtracted from the waveform (*b*), the blue signal. These being voltages, negative values are possible and indeed occur—see (*f*). Similarly, the R − Y signal waveform is shown in (*g*).

G − Y, which is not transmitted but recovered in a matrix at the receiver, is also shown (*h*).

The dotted lines of (*f*) and (*g*) represent the 'weighted' values of the U and V signals applied to the modulators.

To summarise this very important chapter the block diagram of Fig. 10.28 reveals the PAL encoding and decoding actions from the transmitter to the receiver in terms of signal waveforms and vector 'phasors'.

At the top left-hand corner of the diagram is the colour television camera scanning a test card bearing the primary and complementary hues of almost maximum saturation, and yielding primary colour signals of unity value, as shown by the red, green and blue waveforms at the R, G and B camera outputs.

These signals are applied to a network which adds them in the proportions red 30 per cent, green 59 per cent and blue 11 per cent, giving the Y signal, also of unity value.

The red and blue primary colour signals are then fed separately to further networks which subtract the Y signal from each one (this is achieved by the 'box' labelled $-Y$). These same networks also apply the PAL weighting, giving the U and V signals from the original $B - Y$ and $R - Y$ signals, as shown.

Thus, we are left with three signals, the Y signal, the U signal and the V signal, the waveforms of which are illustrated. Notice that the stepped nature of the Y waveform when the colour bars are scanned. It is this 'stepping' which yields the monochrome greys from the colour input.

The Y signal is fed into the composite adder at the top right-hand side of the diagram, after first passing through a delay network, which ensures that the luminance waveform lines up exactly with the chroma waveform shown directly below it. The chroma signal at that point is created by the U and V signals first being applied to the corresponding modulators along with the 4·43 MHz subcarrier signal, obtained from the 4·43 MHz oscillator. It will be seen that the subcarrier is applied to the U modulator direct and to the V modulator (that which is 'alternated') via the 90 degree phase shift network *and* the PAL switch. The 90 degree phase shift network provides the normal conditions required for quadrature modulation, while the PAL switch alternates the phase of the subcarrier fed to the V modulator (derived, remember, from the $R - Y$ signal). It is this action which alternates the phase of the V signal in the manner described earlier in this chapter.

Colour bursts are also applied to the U and V modulators, and these are locked in frequency and phase to the subcarrier signal, for the colour bursts are an extension, so to speak, of the subcarrier for the accurate phasing of the reference generator in the receiver.

The modulators effectively suppress the subcarrier frequency and deliver the U and V sidebands separately. These are shown as 'signal envelopes' at the output of each modulator. When the colour information is zero (colourless input at the camera), the U and V envelopes also shrink to zero, while the amplitude increases as the colour saturation increases. The vector arrows associated with the U and V signals show the constant phase of the U modulation components and the alternated phase of the V counterparts.

Now, the U and V components are integrated in the chroma 'adder', giving the chroma output, including the swinging bursts. The bursts are caused to

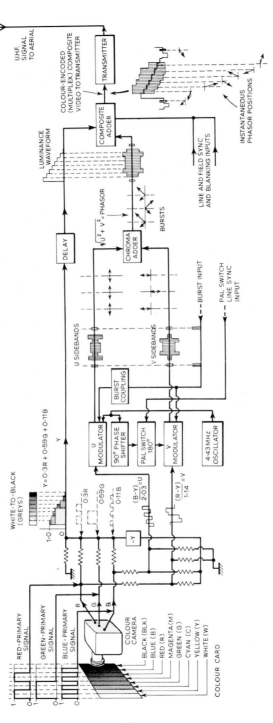

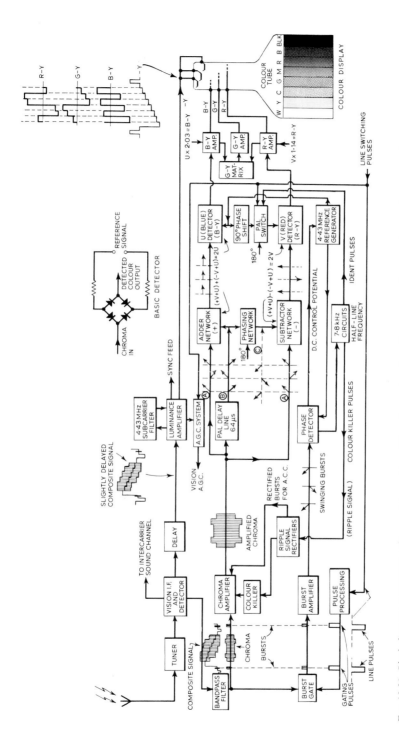

FIG. 10.28. *Block diagram summarising the complete working of the colour system from the camera at the transmitter to the picture tube at the receiver. The signal waveforms and their phasors are revealed at the various points.*

209

swing ± 45 degrees in exact synchronism with the V signal alternations, for it will be recalled that the swinging bursts also synchronise the PAL switching at the receiver.

In the PAL system, therefore, the bursts have the dual function of phase-locking the reference generator and synchronising the V (or R — Y) detector switching. In the NTSC system the bursts are concerned solely with phase-locking the reference generator. As in the PAL system, though, their presence on the signal allows for the removal of the colour-killer bias and for automatic chroma control (e.g. colour a.g.c.) on colour transmissions.

The chroma signal can at this stage be considered as a vector 'phasor' rotating within the colour diagram and lining up instant-by-instant with the scanned elements of the colour scene. The phasor is indicated in the diagram at the output of the adder, as also is that corresponding to the swinging bursts. The phase-reversal action is here shown line-by-line. Notice, too, that the *amplitude* of the phasor at any instant is equal to the vector sum of the U and V signals, this corresponding to saturation, while the relative phase of the phasor corresponds to colour.

The chroma signal applied to the input of the composite adder, still minus a subcarrier and representative of the integrated U and V sidebands, has a waveform based on the addition of the U and V waveforms. At this stage it can be seen how the Y delay network causes the luminance waveform to line up with the chroma waveform, which is essential for correct luminance and colour registration. Delay is required because the bandwidth of the Y channel is about five times as great as that of the chroma channels, and signals pass more speedily through a channel with a wider bandwidth.

The composite adder is concerned with the production of the composite encoded modulation signal, consisting of the chroma information, the Y signal, the colour bursts, the line and field sync pulses and the blanking (black level) pulses, so as to resolve the signal shown in Fig. 10.26. One line of composite (multiplex) modulation signal is shown at the output of the composite adder, along with the associated phasors. Bursts corresponding to *two* lines are shown so as to reveal the phase alternations—line-to-line. The approximate phase of the chroma signal is shown from colour-to-colour by the phasors, and the colours represented by the waveform sections can be identified in Fig. 10.26. Note, though, that the composite waveform on the diagram is an inverted, mirror image of that in Fig. 10.26, because the former represents *modulation signal*, and the latter modulated *carrier signal*.

Finally, the colour-encoded (multiplex) composite video signal is applied to the modulators of the transmitter and propagated through the ether in the ordinary manner.

We now come to the bottom section of the block diagram which represents a single-standard, 625-line receiver. The complications of dual-standard working are unnecessary for the purpose of this description, and in any case most receivers in a few years time will be single-standard versions.

The carrier signal is picked up by the aerial, selected by the tuner, amplified at i.f. and finally detected. The detector yields a replica of the modulation signal at its output, which is fed in two directions—(1) to the chroma amplifier, and (2) to the luminance amplifier—via a Y-channel delay (not to be confused with the PAL delay line), corresponding to that at the transmitter, with a similar purpose. The composite signal at these two points in the diagram is shown. Let us follow the Y signal or luminance path first. On arriving at the luminance amplifier, a 4·43 MHz filter removes the chroma signal, leaving just the stepped waveform of the Y signal. This is inverted by the action of the luminance amplifier and thus arrives at the cathodes of the picture tube as $-$Y signal, as shown. The luminance channel is, in fact, almost identical to the video channel in ordinary monochrome sets.

In the chroma channel, the composite signal is passed into the chroma amplifier through a bandpass filter which removes all the Y information and sometimes the bursts, leaving at the output only the chroma signal. This signal is amplified and appears at much larger amplitude at the chroma output, and at this point it is fed into the PAL delay line and the associated adding and subtracting network. It is interesting to note that the chroma signal at this point is the same as that at the output of the chroma adder in the transmitter.

It will be recalled that the output of the PAL delay line represents chroma information from the preceding line of signal, in which the V signal has opposite phase; thus, the outputs from the adder and subtractor networks correspond respectively to the addition and subtraction of adjacent lines of colour information, thereby resulting in phase-correct signal which is passed on to the U and V detectors.

In the diagram, lines (A) from the chroma output feed the signal direct to the adder and subtractor networks, while line (B) feeds the delayed signal into both networks—to the subtractor, via a network which provides the necessary phasing.

Let us suppose that the signal at the output of the PAL delay line is $+$ V $+$ U, which would mean that at the same instant the signal direct from the chroma amplifier—corresponding to the next line—would be $-$ V $+$ U. The adder network adds these two signals to give

$$(+ \text{ V} + \text{U}) + (- \text{ V} + \text{U}) = 2\text{U}$$

while the subtractor networks subtract them to give

$$(+ \text{ V} + \text{U}) - (- \text{ V} + \text{U}) = 2\text{V}$$

The effect is that the U components are cancelled out and that the 2V component remaining reverses in phase line by line, so that $+$ 2V is given on one line, $-$ 2V on the next line, and so on.

It is in this way, as was shown earlier with *colour-difference signals*, that the adder and subtractor networks cancel out phase errors while providing separate V and U signals, which are free from phase distortion, for the colour detectors.

The phasors associated with lines (A) on the diagram represent the phase angle of the chroma signal towards the red part of the colour scene. Those associated with line (B) represent the reverse phase angle of that same part of the chroma signal on the preceding line of picture applied to the adder. The phasors associated with line (C) represent the phase-reversed delayed signals applied to the subtractor.

The phasors on the lines from the outputs of the adder and subtractor networks represent the phase-corrected and 90 degree displaced U and V chroma signals (V signal alternating) as applied to the U and V detectors. It should be remembered that the nominal phase of these signals remains constant in spite of the signal being put in phase error as it passes from the input of the transmitter to the output of the receiver, via the ether. Any phase error resulting in the system simply alters the lengths of the phasors, and hence the colour saturation—not the hue.

The colour detectors can be considered as diode switches operated by the 4·43 MHz reference signal, delivered by the reference generator. A four-diode bridge circuit is often used in each detector, as shown in the middle of the diagram, the action of which is as follows. When the reference signal swings in polarity such that the diodes are rendered conducting (e.g. positive at the top terminal and negative at the bottom one), the chroma signal fed to one side of the bridge passes through the conducting diodes and appears as 'detected' V or U signals at the other side of the bridge. Of course, when the reference signal swings such that the diodes are rendered non-conducting the chroma signal is 'blocked'.

The time constant of the circuit is arranged so that the diodes conduct only on the peaks of the reference signal, and in this way the chroma waveform is sampled at intervals corresponding to the frequency of the reference generator.

It will be recalled that the reference signal must be exactly phased to the suppressed subcarrier. When this is so the detector output matches the modulation applied to the corresponding modulator at the transmitter. However, because the reference signal has a 90 degree phase difference between the two modulators and detectors, the 'blue' detector produces an output corresponding only to the 'blue' modulation and the 'red' detector produces an output corresponding only to the 'red' modulation. In other words, quadrature modulation allows two 'bits' of information to be carried simultaneously in a single channel while providing the means for separating the information at the receiver.

This is the reason, then, why it is so important for the phase of the reference signal at the receiver exactly to match that of the suppressed subcarrier at the transmitter. It is also the reason why the chroma information can be carried along with the Y signal in a channel of 'standard' 625-line width.

The reference signal is fed to the U detector direct and to the V detector via the 90 degree phase shift network and the PAL switch. These items correspond to those at the transmitter, the 90 degree phase shift providing the normal conditions required for quadrature demodulation, and the PAL switch alternating the phase of the V signal line-by-line by means of line pulses picked up from the line timebase. It is, of course, essential for the V detector to switch phase to correspond to the PAL V signal alternations.

From the chroma signal input end of the diagram (bottom left) it can be seen that the chroma signal is also applied to the burst gate. The burst gate is switched to open by processed line pulses and, therefore, only lets through the bursts, as shown at the output of the gate. The bursts are amplified and then fed to the phase detector along with a sample reference signal. Should the phase of the reference signal tend to drift from that of the bursts, the phase detector delivers a potential which adjusts the reference generator accordingly and restores the correct phasing.

Owing to the swinging bursts the phase detector also delivers a ripple signal, equal to half the line frequency (7·8 kHz), which is used to identify the phase-reversed lines of V signal by adjusting the PAL switch to the correct 'count', as already described.

Any change in amplitude of the chroma signal will automatically change the amplitude of the burst signal, and because of this it is often rectified to produce the chroma a.g.c. bias, for giving automatic gain of the chroma channel (automatic chroma control—a.c.c. for short). Moreover, since the burst or ripple signal exists only when colour information is present on the signal, it is often used to operate the colour killer; that is, it brings the chroma amplifier into action during a colour transmission and automatically mutes it when the transmission is in monochrome. In this way it ensures that spurious colour effects are not encouraged—due to interference and noise signals etc.—on black-and-white pictures.

Finally, we must again turn to the colour detector end of the diagram and see that the U and V outputs are fed to the corresponding B − Y and R − Y amplifiers, adjusted to provide the correct relative gains to compensate for the weighted signal (in other words, changing the U and V signals back to the original B − Y and R − Y signals).

The G − Y signal is obtained by matrixing correct proportions of the R − Y and B − Y signals (as shown earlier). The amplified colour-difference signals are applied to the corresponding grids of the picture tube, with −Y at the cathodes deleting the Y content. The guns, therefore, 'see' only the primary-colour signals at the R, G and B outputs of the camera, from whence we started our journey through the colour system.

The picture tube thus displays colour bars which match very closely those scanned by the camera. The white bar contains a mix of R, G and B lights, the yellow bar contains a mix of G and R, the cyan bar contains a mix of G and B, the green bar contains G only, the magenta bar contains a mix of R and B, the red bar contains R only, the blue bar contains B only, and the black bar has zero light—all phosphors unlit.

It should be borne in mind that some sets might differ in detail from the block diagram shown. The V and U compensation, for example, may be built into the adder and subtractor networks, which would mean that the colour detectors would then be true colour-difference ones, as distinct from U and V counterparts. Alternatively, some of the compensation may be introduced at the adder and subtractor networks and some after the colour detectors, in the colour-difference amplifiers.

INTEGRATED CIRCUITS

Integrated circuits are finding their way into colour receivers, and an interesting circuit in this respect is the i.c. colour decoder used in some of the RBM receivers. This is based on a 20-lead encapsulation designed to replace 65 discrete components in the stages shown in Fig. 10.29.

This shows that the circuit not only yields internally the G − Y signal by matrixing the B − Y and R − Y signals, but it also gives red, green and blue primary colour signals. The circuit receives three inputs, Y, V and U, and delivers R, G and B outputs of suitable level for driving three primary colour

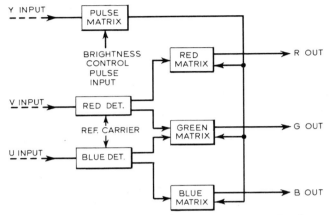

Fig. 10.29. *Block diagram showing the functions performed by the RBM decoder integrated circuit.*

amplifiers. Brightness is regulated by a pulse circuit in the Y channel, this being arranged to regulate the biasing on the tube guns, via the matrices and the d.c. couplings therein.

The chroma detectors are different from the diode 'envelope detectors' so far considered in that a system of seven transistor elements (part of the i.c.) is employed in the form of a pair of differential amplifiers, giving balanced detection. Y and colour-difference matrixing (to obtain the primary colour signals) employs three transistor elements (still in the same i.c.) in the form of a 'level shifting' amplifier, one for each colour. The operation of this is similar to that in the BRC sets, but these use separate power transistors. The primary colour outputs of the RBM i.c. feed into primary colour amplifiers and then to the guns of the picture tube.

Integrated circuits are also being employed in the sound channel and sometimes in the intercarrier channel as well, with a quadrature demodulation. In this a three-transistor-element section in the i.c. is arranged rather like a differential amplifier, with the frequency modulated signal being applied to both inputs, but the phase to one input shifted. Audio signal is then delivered at the collector of one of the transistor elements which is loaded resistively.

The large device firms like Mullard are making i.c.s for colour sets (and for monochrome ones), an example being the Mullard TAA570. This provides sound demodulation and low-level audio-frequency amplification. An ordinary diode is employed as the intercarrier detector (to obtain the 6 MHz frequency-modulated signal) and this is fed into the i.c. which itself features a quadrature detector for deriving the audio signal.

CHAPTER 11

Test Instruments and Signals

MANY OF THE test instruments employed for radio, audio and monochrome television servicing are also required for the servicing of colour sets. However, for the successful servicing of colour sets, instruments which might be considered unnecessarily elaborate or specialised for the servicing of less complex equipment are often essential. These special instruments are designed essentially for colour set adjustment, maintenance and servicing. They particularly facilitate adjustments in, and the servicing of, e.h.t. circuits, convergence circuits, grey-scale tracking circuits and the various circuits associated with the PAL decoder, while most of the existing test instruments come into use for the less involved activities associated with the basically monochrome sections of the colour receiver.

BASIC INSTRUMENTS

Basic 'non-colour' test instruments thus include a multirange testmeter which should preferably have not less than 20,000 Ω/V sensitivity and full-scale ranges suitable for measurements in the relatively low voltage transistorised circuits, an r.f. generator for tuner and i.f. circuit alignment which extend to the v.h.f. TV bands though not necessarily rising into the u.h.f. bands, and a wobbulator and oscilloscope partnership for visual alignment.

INSTRUMENTS FOR COLOUR SERVICING

There are six prime operations in colour set servicing (including adjustments) for which instruments of some kind are necessary. They are: (*i*) convergence (static and dynamic), (*ii*) purity, (*iii*) grey-scale tracking, (*iv*) e.h.t. voltage adjustment and measurement, (*v*) degaussing and (*vi*) decoder adjustments.

Operations (*i*), (*ii*) and possibly (*iii*) can be carried out by using the signals provided by a *crosshatch and dot generator*. Operation (*iv*) requires either an *e.h.t. voltmeter* or a *voltage multiplier* for extending the voltage range of a 20,000 Ω/V voltmeter to measure up to 25 kV or 30 kV. Operation (*v*) is partly carried out by the set's auto-degaussing system, but an *external degausser* is desirable when first setting up a colour set and for demagnetising large metal items near to the set. Operation (*vi*) can be handled by a *colour-bar generator* in conjunction with an *oscilloscope* and/or *valve voltmeter*.

The colour-bar generator is essentially a colour-encoded signal source, matching the PAL and/or NTSC standards (switchable on some models), for which reason it yields a stepped monochrome signal (with the chroma switched off), ideal for grey-scale tracking. Most such instruments also deliver cross-hatch and dot-pattern forming signals for convergence and linearity adjustments.

INSTRUMENT SURVEY

It will now be instructive to consider the various instruments required for efficient and speedy colour set servicing in a little more detail.

Crosshatch and Dot Generator

This is a relatively simple instrument, having much in common with a mono-chrome television pattern generator. There are portable battery powered and transportable mains powered versions, both of which at the time of writing are mostly designed for dual-standard operation. The modulation signal, producing

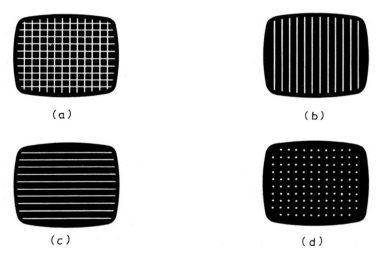

(a) (b) (c) (d)

FIG. 11.1. *Patterns from a crosshatch and dot generator. (a) crosshatch, (b) vertical lines, (c) horizontal lines and (d) dots (these normally occur at the intersections of the crosshatch lines).*

the crosshatch and dot patterns on the television screen, also contains line and field sync and blanking pulses, and the composite signal is modulated on to an r.f. carrier wave, which is often tunable over the v.h.f. and u.h.f. television channels. In some models the modulation signal is available at video frequency from a separate outlet.

Different models exhibit variations in detail, of course, but it is not uncommon to find controls for regulating the number of vertical and horizontal lines of the display, switching off the modulation—thereby giving an unmodulated raster upon which purity adjustments can be performed, selecting the required 405-line or 625-line standard, switching out the vertical or horizontal lines—leaving only the remaining partners of the crosshatch, reversing the signal polarity and for adjusting the composite signal amplitude and the amplitude of the sync pulses. Some models also give a grey-scale pattern.

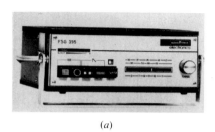   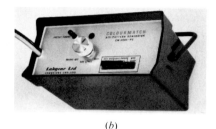

(a)   (b)

Fig. 11.2. *Examples of crosshatch and dot generators. (a) Normende FSG 395, distributed in the U.K. and abroad by Bemex Instruments Ltd. (b) Model CM 6004-PG (Labgear Ltd.).*

Figure 11.1 shows some of the patterns that can be provided by a crosshatch and dot generator. With signal polarity reversing facilities, it is possible either to have white patterns on a grey background or black patterns on a white background. Examples of the simple crosshatch and dot generator are illustrated at (a) and (b) in Fig. 11.2.

E.H.T. Voltmeter and Multiplier

As the application of e.h.t. voltage fed to the colour tubes is rather critical, and as excessive voltage can produce x-rays above the safe dose-rate, some means of determining the e.h.t. voltage is essential. There are specially developed e.h.t. voltmeters reading to 25 kV and above, but many servicing technicians prefer either to calculate the e.h.t. voltage by measuring a relatively small voltage across a resistor carrying the e.h.t. regulator current on a high resistance voltmeter, or to employ an e.h.t. multiplier for use with an existing 20,000 Ω/V voltmeter.

Such a multiplier is generally built into a highly insulated test probe, such as shown in Fig. 11.3(a) which is designed for use with an instrument of the kind shown in Fig. 11.3(b).

External Degausser

An external degausser is formed either of a coil-loop or solenoid designed for intermittent mains operation. An example of the coil-loop version is shown in

Fig. 11.4. This is equipped with a length of three-core mains cable and a 'hold-on' push-button switch which makes it impossible to leave the instrument inadvertently switched-on—an action that would quickly bring the coil to a very high temperature with a possibility of subsequent insulation collapse.

It will be understood, of course, that a high intensity a.c. magnetic field is necessary to remove residual magnetism from a large metal object, for which

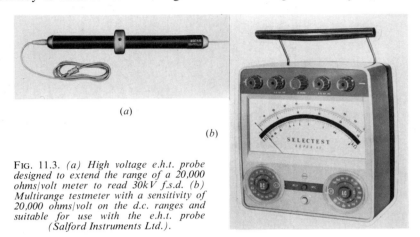

(a)

(b)

FIG. 11.3. (a) High voltage e.h.t. probe designed to extend the range of a 20,000 ohms/volt meter to read 30kV f.s.d. (b) Multirange testmeter with a sensitivity of 20,000 ohms/volt on the d.c. ranges and suitable for use with the e.h.t. probe (Salford Instruments Ltd.).

reason a high current must flow through the coil winding. Thus, to give the instrument practical dimensions, the coil can only be rated on an intermittent basis. For continuous operation, which is not necessary anyway, a degausser would need to be many times larger than that pictured in Fig. 11.4. Details of how to demagnetise with an external degausser are given in Chapter 6.

Colour-Encoded Generator

Some generators in this category provide 'standard' colour bars—such as those transmitted by the BBC (and at the top of Test Card F)—in addition to the various patterns portrayed in Fig. 11.1. Of course, for British and European applications, the generator must be encoded to the PAL parameters, including the swinging colour burst. Some models can be switched to give PAL and NTSC signals, an action which not only modifies the chroma weighting factors and the burst phase relative to the colour axes, but which also changes the burst amplitude and introduces a swinging burst in the PAL position.

It is not uncommon to find signal outputs at r.f. (tunable over the u.h.f.—and sometimes v.h.f.—bands) and video, the latter allowing the colour-encoded signal to be applied direct to the decoder or video detector output, as may be required for the tests and adjustments. Outputs may also be featured for the burst signal alone, and for the line sync pulse signal, which enables an associated oscilloscope to be 'locked' to the actual composite signal applied to the set for testing.

The composite signal, in fact, might be almost a replica of that transmitted by the authority when televising the colour bars (see Fig. 10.26). Clearly, then, such an instrument is bound to be much more costly than the far less complex crosshatch and dot generator which is devoid of colour-encoding circuits.

Some colour generators, while providing colour bars and a rainbow display in addition, are not designed to yield 'standard' signals; but the signals that they do produce are perfectly adequate for evaluating the set's colour performance and for decoder adjustments.

Whatever the nature of the colour generator, the circuitry is extremely involved, equal at least to that of the colour receiver itself, and most instruments nowadays are fully transistorised, see Fig. 11.5(c).

Instruments with switchable PAL/NTSC facilities can exploit the *NTSC* signal for various tests and adjustments in *PAL* sets, and the value of the instrument is enhanced if it is possible to control the amplitude of the colour-burst signal from zero upwards.

Owing to the stepped luminance waveform derived from the colour-bar signal (see Fig. 10.28), this modulation can be used to produce the typical grey-shade

FIG. 11.4. *Example of a degausser (Labgear Ltd.).*

bars on a monochrome set and on a colour set with either the chroma switched off at the generator or at the set itself (e.g., with the saturation control turned right down). This pattern, then, is ideal for grey-scale tracking adjustments and assessments. When the burst source is switchable at the generator, the chroma signal automatically disappears from the modulation and leaves only the stepped luminance waveform, as just mentioned. Thus, by observing the grey-shade bars initially on monochrome, the effect of the chroma signal on the display immediately becomes apparent by switching on the burst signal source, revealing how well the various rejector circuits perform.

Plate 2 shows at (a) the standard stepped waveform display on monochrome, at (b) the standard colour-bar display with the chroma switched on. The colours at the top of Test Card F can be useful in the absence of a colour bar transmission and exposure is possible either by slipping the field or by shifting the frame down the screen slightly as shown respectively at (c) and (d) in Plate 2.

Colour generators capable of yielding a diversity of displays are also available. The standard colour bars can be obtained from many of them and some also provide special colour-encoded signals to facilitate adjustments in the PAL decoder. In addition to the Körting colour generator referred to elsewhere, another recent example is the Philips PM 5508. This solid-state instrument is

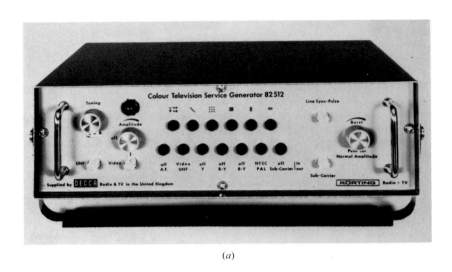

(a)

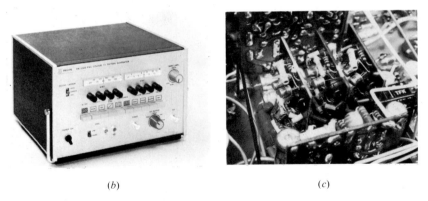

(b)   (c)

FIG. 11.5. *Examples of colour-encoded generators. (a) Model 82512 Körting, distributed in Great Britain by Decca (courtesy, Decca Radio and Television Ltd.), and (b) Model PM5508 Philips (courtesy, Pye-Unicam Ltd.). The solid state modules of (a) are shown in (c).*

equipped with push-button test selection and produces signals suitable for making accurate adjustments in the circuits of the PAL delay line and matrix (phase and amplitude) and in the synchronous detectors using either the displays on the television screen or the traces on an oscilloscope (or both). Examples of colour pattern generators are shown in Fig. 11.5.

Oscilloscope

Generally speaking, any oscilloscope with a Y response extending up to 4·5 MHz is suitable for colour servicing. A level response over this bandwidth is necessary, of course, to avoid undue attenuation of the burst and chroma signals, for these often have to be examined either separately or as a part of the composite colour signal. To ensure that the amplitude of the bursts relative, say, to the line sync pulses is reasonably accurately indicated, the Y bandwidth should be pretty flat up to at least 4·5 MHz. It might also help in this connection to employ a low-capacitance probe in the Y-input circuit of the oscilloscope. An ordinary 'service type' oscilloscope can be used in most cases, but if the specification indicates that its Y response extends to 4·5 MHz or thereabouts, it is as well to make certain that the amplification at the top frequency is not too many decibels down! On the basis of an overall Y response, therefore, a value approaching 10 MHz would be a safer bet, for such an instrument would almost certainly have full amplification at 4·5 MHz.

The larger the diameter of the tube face the better, of course, but a screen of about 8 cm diameter is perfectly adequate for most applications. A green trace is the least fatiguing when extensive use is made of the oscilloscope for colour television servicing, which can often be a protracted exercise!

The sweep (X) velocity should match the Y bandwidth. This allows the separate waves of the burst signal to be resolved should this ever be necessary to check wave-shape, etc. Thus, the maximum sweep should not be less than about 1 μs/cm which, with X expansion, easily provides the requirement.

Internal and external repetitive and trigger sync facilities with polarity reversing are desirable, bearing in mind that it might be required to 'lock' the trace to the line sync pulses produced by a pattern or colour generator. Another good facility is grid modulation—called Z modulation.

Valve Voltmeter

Low voltage measurements often have to be taken from high resistance circuits in colour sets, especially around the decoder section, and to avoid damping or altering the characteristics of the circuits the testmeter must impose the very least shunt resistance (or impedance). Many such tests and measurements can be performed with a voltmeter having a sensitivity of not less than 20,000Ω/V, while the 100,000Ω/V instrument is even better, almost as good as some valve voltmeters, especially on the higher voltage ranges.

Nevertheless, there is a good case for the valve (or equivalent transistor) voltmeter or multirange electronic testmeter when small voltages need to be measured in high resistance circuits. This kind of instrument can also be used with an e.h.t. multiplier (of a suitable matching type, of course) to yield a f.s.d. of 25 kV or so.

An important point to remember when selecting a testmeter these days is that more and more television receivers (including colour ones) are changing from valves to transistors, via the hybrid designs. The smaller transistor voltages must, therefore, be catered for on the testmeter—not by a deflection over a small fraction of the scale, but by at least a half-scale deflection.

The above mentioned instruments represent those needed specifically for colour set servicing. Some of them will already be at hand and very much earning their keep for radio, audio and monochrome equipment servicing. It might only be necessary, therefore, to purchase a degausser, a generator of some kind, and an e.h.t. multiplier. It is possible to look upon colour instruments as those required for the initial setting-up of a colour set in the home and those required for fault-finding and servicing.

Field instruments must certainly include the crosshatch and dot generator (if not the colour-encoded generator), the degausser and, possibly, the e.h.t. test-meter (see Chapter 14).

Whether or not one decides to invest in an expensive colour-encoded generator will depend on the prevailing economics and on how many colour sets are expected to be serviced during the initial colour build-up period. It is possible to perform many servicing operations on colour sets without the colour-encoded generator, using the colour bars and Test Card F, for instance. Where relatively large numbers of sets are likely to pass through the service department, however, the colour-encoded generator can certainly prove to be a very good investment.

OTHER INSTRUMENTS FOR COLOUR

Let us now look briefly at some of the instruments of secondary importance required for servicing the new-generation colour sets. One of the most useful instruments in this category—for monochrome too—is the *signal strength meter*. This is essentially an r.f. voltmeter which is tunable over the v.h.f. and u.h.f. bands and designed to record the very small signal voltages that generally exist at the end of the aerial downlead. An instrument having a full-scale deflection corresponding to 1 or 2 mV of signal with an input attenuator or switch position multiplying the range by 10 and/or 100 times is desirable.

This kind of instrument is of secondary importance at least, because a colour set demands a fairly strong and consistent aerial signal to yield good colour pictures. Indeed, it has been computed that (other things being equal) a colour set needs about 50 per cent more aerial signal than a monochrome counterpart to produce pictures of subjective equivalence from the noise point of view. Thus, while a monochrome set may give pictures at a subjective noise threshold on, say, Channel 24 with an aerial signal of 1 mV, a colour set using a tuner of similar noise performance would need a signal of about 1·5 mV to produce a colour picture carrying about the same amount of noise. In other words, an input of about 1 mV applied to such a colour set would result in rather a noisy picture (e.g. colour grain).

In many areas, therefore, this calls for a u.h.f. aerial of high gain and directivity, and the only true way of ensuring its accurate orientation is to measure the signal at the end of the feeder with a signal strength meter. The signal strength meter also makes it possible to decide conclusively whether poor reception (colour noise, etc.) is due to a weak aerial signal or a fault in the set.

Another useful instrument is that which allows the screen of the colour picture tube to be conveniently viewed whilst making adjustments at the rear of the set. In its simplest form, of course, this is just a clear reflecting surface, such as a household mirror. An improvement on this is a shaving mirror with a concave reflecting surface on one side which can be used to magnify a section of the screen for close scrutiny of the dot phosphors and for accurate convergence adjustments.

An interesting commercial evolution of the magnifying mirror is the Beam Landing Pattern Scope made by Philips. This instrument, shown in action in Fig. 11.6, consists basically of a periscope and microscope combined and enables the technician to view the image on the screen from the rear of the set, so

FIG. 11.6. *Beam Landing Pattern Scope by Philips (courtesy, Pye-Unicam Ltd.). This allows a technician to view the tube face in detail from the rear of the set (see text). Plate 3 shows how the illuminated dots are seen.*

eliminating the need for repeated movements between the screen and the controls. The magnification is about 30 times, and it has built-in battery-powered lighting. The height can be adjusted. Plate 3 shows the magnified, illuminated dots on a modulated raster.

Adjustable, arm-mounted mirrors are also available for rear-of-set screen viewing, but for dynamic convergence adjustments the convergence adjustment panel of most sets can be detached and brought to the front of the set on an extended cable.

An audio/video signal generator delivering sine and/or square waves is another useful item of equipment, allowing response tests to be made in the chroma and video channels (luminance and colour-difference).

A 'white' reference provided by a fluorescent tube, enabling the correct illuminant D to be established easily by adjusting the receiver's controls, is a recent instrument by Blocktube Controls Ltd.

MONOCHROME INSTRUMENTS REQUIRED FOR COLOUR SERVICING

Colour set servicing also calls for the use of some instruments used for monochrome set servicing. A major instrument in this category is the r.f. modulated signal generator, essential for tuned circuit alignment in conjunction with sound and video output meters of some kind. However, with the added complexities of dual-standard i.f. strips and the greater accuracy of response required for colour sets, the visual alignment technique using an r.f. generator, wobbulator and oscilloscope, is often recommended. Some manufacturers' service manuals cover the alignment technique by both methods, while others detail only the 'visual' method.

The oscilloscope should feature an X output terminal for sweeping the wobbulator if necessary, but otherwise it can have similar features to those mentioned earlier. Any r.f. generator is suitable for putting a frequency marker pip on the displayed response curve. Crystal-controlled marker generators, or an inbuilt crystal oscillator giving harmonics to check the r.f. generator tuning, can be useful when it comes to adjusting the various rejector circuits and the intercarrier sound channel (at 6 MHz) which require great accuracy. Some r.f. generators, such as the Grundig WS3, incorporate 50 Hz sweep-frequency facilities from zero to 30 MHz deviation (adjustable) with r.f. signal output from 4 to 230 MHz (v.h.f.) and from 470 to 800 MHz (u.h.f.), with inbuilt marker oscillators and crystal calibration.

This sort of instrument will handle all alignment problems—v.h.f. and u.h.f. tuners, i.f. channels, intercarrier sound channel and rejectors—using either the point-frequency and output meter method or the visual method with a partnering oscilloscope.

COLOUR SET CIRCUIT ALIGNMENT

Although some technicians with many years of experience in monochrome servicing are able to align black-and-white sets by ear and eye, so to speak, accurate alignment of colour sets, ensuring the best colour performance, is possible only by closely following the instructions given by the particular manufacturer in his service manual for the particular model. Of course, the basic scheme of things has much in common with the alignment of monochrome sets, but with colour the many rejectors have to be very carefully aligned at their exact frequencies. The vision i.f. channel response must be free from abnormal undulations and the sound, vision and colour carrier frequencies must be accurately placed at specified levels along the overall response curve. Unless this high degree of alignment accuracy is achieved, the colour performance will fail to reach the available standard and disconcerting pattern effects will tend to spoil the pictures.

TRANSMITTED TEST SIGNALS

A useful colour test display is provided by the standard colour bars (see Plates 2(b) and (d)). This can be employed to evaluate the colour performance of an installed receiver, and can be used in a limited way to make adjustments inside the set (see Chapter 13). The line waveform of this signal is given in Fig. 10.26, while the signal itself is fully investigated in Chapter 10. However, so far there has been no reference to the field synchronising signals, the burst signals during the field sync period and the burst blanking.

While the burst signals of the NTSC system are suppressed during the field sync period, the phase of the swinging bursts of the PAL system starting and ending the fields when similarly suppressed can give rise to complications with

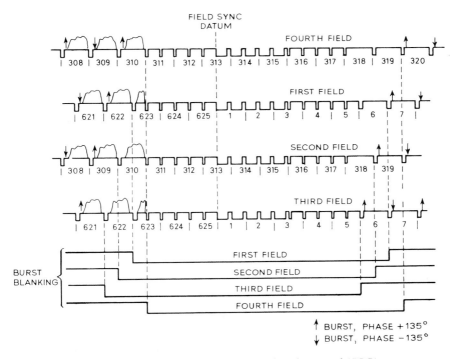

FIG. 11.7. *Field sync and burst blanking of a colour signal (BBC).*

certain types of receiver reference generators. For instance, the starting phase after the field interval may fail to coincide with that ending the field. This condition can incite a slight flicker effect at the top of the picture when the colour saturation is high.

This is combated in the PAL system by the employment of a special burst blanking sequence which ensures that all the fields start and finish with the same burst phase. This can be better understood by studying the BBC's field sync signals shown in Fig. 11.7.

COLOUR TEST CARD F

Test Card F is the colour equivalent of the monochrome test cards, such as Test Card C and Test Card D, and is shown in Plate 6. It was designed jointly by the BBC, BREMA, EEA and the ITA, and is significantly different from the monochrome counterparts in that the edge castellations are coloured, as also is the picture contained within the centre white circle. Apart from this colour information (described later) the card possesses numerous features which are common to the monochrome test cards. These include a background grid with corner diagonals and a centre circle for appraisal of picture geometry, frequency gratings designed to produce signals of approximately square-wave form corresponding to frequencies of 1·5, 2·5, 3·5, 4·0, 4·5 and 5·25 MHz for appraisal of the picture definition, a 'letter box' pattern to reveal low-frequency effects and a six-step grey scale enabling the performance of the video circuits (contrast) to be assessed. The card thus provides tests for the various functions of both colour and monochrome sets.

Let us first investigate the basically monochrome tests, which are as follows.

Uniformity of Focus

This is appraised by how well the set focuses the areas of black and white diagonal stripes in each corner of the card.

Reflections

Signal reflections from hills and large buildings produce displaced 'ghost' images, called multipath interference, which can be readily observed as displaced 'ghosts' of the white and black vertical lines, particularly where they are adjacent.

Low-frequency Response

This can be assessed by the black rectangle within the white rectangle at the top of the card. Impaired l.f. response is revealed by streaking at the right-hand edges of these areas and border castellations.

Line Synchronisation

On monochrome sets the black and coloured rectangles bordering the card appear as black and various tones of grey down to white. Faulty line sync shows as horizontal displacement of those parts of the card down the *right-hand* border which are on the same level as the less grey rectangles. A by-product is the formation of a 'cog-wheel' effect on the centre white ring.

Scan Linearity

This is appraised by how well the centre white ring forms a circle and the background grid forms squares.

Resolution and Bandwidth

The receiver vision bandwidth corresponds roughly to the frequency labelling the highest definable frequency grating. For example, the bandwidth would be roughly 4·5 MHz if all the gratings up to and including the 4·5 MHz grating were well defined. In this case, of course, the 5·25 MHz grating would appear as a grey rectangle, rather than a series of closely spaced vertical stripes.

Contrast Performance

This is revealed by the gradation displays of the column of six rectangles located to the left of the centre circle. The transmission has a contrast ratio of about 30:1. On a correctly designed and adjusted set the brightness difference between adjacent rectangles should be reasonably constant. The small, lighter-shade spots in the top and bottom rectangles show white or black contrast crushing respectively by the spot tending to merge into its surrounding area.

Picture Size

The card is a little smaller than an actual transmitted picture, the width and height limits of which are indicated by the points of the opposing arrowheads in the borders. Some sets have a tube display area corresponding to an aspect (width-to-height) ratio of about 5:4, so it is usual to adjust the height control of the set first until the top and bottom edges of the screen just coincide with the arrowheads. The picture width is adjusted afterwards to provide the correct display geometry, and this will cause the side castellations of the card just about to fit across the screen (assuming, of course, that the picture is properly straightened and centred).

It is noteworthy that the more recent tubes have an aspect ratio close to 4:3, matching that transited (see below), so the compromise height/width adjustments will not apply so much to sets using these.

Aspect Ratio (Width-to-Height Ratio)

The standard transmitted width-to-height ratio is 4:3 (as opposed to the 5:4 aspect ratio of some tubes), and this is automatically achieved when the width and height of the picture are adjusted to yield a truly circular centre white ring.

COLOUR TESTS

The various coloured parts of the card are designed for judging the colour performance of the receiver as follows.

Decoder Performance Check

The first four lines of each field scan are electronically modulated with the standard colour-bar signal of 100 per cent amplitude and 95 per cent saturation*. Thus, the top of the card starts with a short, horizontal strip carrying the standard colours of white, yellow, cyan, green, magenta, red, blue and black. Because they occur four lines after the end of the field-blanking period, the colour-bar signals can be displayed on an oscilloscope by synchronising its timebase to the field scan (see Fig. 10.26) and any irregularity in shape, asymmetry or amplitude of waveform from line to line indicates a shortcoming in the performance of the PAL decoder or colour circuits. In addition, the displayed colours can be used to assess the decoder performance by switching off individual guns of the tube (Chapters 12 and 13).

Reference Generator Faults

How well the reference generator locks to the colour bursts following the field-blanking period can be found by the effect on the top border, high luminance, cyan castellations. Normally, these castellations should be of a constant saturation, but if there is a tendency towards saturation variations, then the reference generator may not be recovering phase and frequency sufficiently quickly, possibly as the result of a phase detector or reference oscillator control shortcoming. Conversely, the bottom, green castellations indicate the performance of the reference generator at the end of the fields, compared with their start.

Burst Gating Faults

If the burst gate circuits allow picture information to pass into the reference generator, then bands of violent saturation change and/or 'Venetian blind' interference (depending on decoder design) will occur across the picture corresponding to the red and blue castellations on the left border of the card, which are included to subject the colour circuit to the greatest disturbance.

Sync Separator Performance

Any malfunctioning of the sync separator circuits gives rise to variations in the position of the picture content on the extreme right, where on the card the

* Some colour test cards, notably of the IBA, use the EBU bars of 75 per cent amplitude and 100 per cent saturation. The percentage may be different on some instrument-derived colour-bar signals and other transmissions.

castellations are yellow and white. This enables the performance with subcarrier (yellow) and without subcarrier (white) to be appraised.

Convergence Appraisal

Some idea of the dynamic convergence can be obtained from the white lines of the background grid which are outlined in black. Some lines have no black out-lining to avoid confusing symptoms of ringing when the set exhibits certain types of quadrature demodulation. An appraisal of static convergence, in the centre of the screen, is provided by the blackboard and white cross in the coloured picture. However, for physical convergence adjustments a crosshatch and dot generator should be used, as the card is not designed as a substitute for this signal.

The Colour Picture

The areas of flesh tones and bright colours in the centre picture allow an overall quality assessment to be carried out without changing from the card and while observing the other factors already discussed. The model of a child rather than an adult is used to avoid changes in fashion of make-up and so forth which would tend to date the card.

The circle enclosing the picture is outlined in white to minimise 'optical illusion' effects which could occur if the colours abutted directly on to the background of the grey and the white grid.

CHAPTER 12

Locating the Fault Area

SUCCESSFUL colour set servicing calls for an awareness of how maladjustment, reduced efficiency and all degrees of failure in any of the circuits leading to the picture tube and speaker can influence the sound and vision performance of the set, both on colour *and* monochrome. Technicians who are used to monochrome servicing will already have this awareness which, for colour, will simply need to be expanded to embrace the additional circuits, which have no monochrome parallel.

It is not intended in this book to explore the various procedures adopted for servicing monochrome sets, but these must be thoroughly understood before it is possible to make a success of servicing colour sets.

A companion to this book is my earlier work, *Television Servicing Handbook* (3rd edition, Newnes-Butterworths Technical Books), which finishes, from the servicing point of view, where this book starts. It deals with most of the problems and procedures relating solely to monochrome set servicing, including single and dual-standard models and the latest transistorised sets; this book is, in fact, written in a manner to complement the earlier work.

As with any other kind of servicing, the initial problem is to locate the area in which the fault lies and then, by more detailed tests, to locate the circuit or component responsible for the trouble, finally clearing the fault by circuit adjustment or component replacement. This chapter is concerned mostly with locating the area in which the fault exists, while Chapter 13 delves more deeply into the tests and adjustments in the fault areas.

COMPLETE FAILURE

A totally dead set—both sound and vision—will almost certainly have a fault in the mains input or power supply circuits. Colour sets run at a greater input power than monochrome counterparts and so to some extent they are more vulnerable to failure of the power carrying components. It is not uncommon to find two power rectifiers (silicon type), each connected to the mains input through its own surge limiting resistor, to handle the greater power demands. In this type of circuit, therefore, failure of one surge limiting resistor or rectifier would not kill the h.t. supply altogether, but it would put a greater load on the active components and result in reduced h.t. voltage with symptoms associated with this kind of trouble. Eventually, the other half of the circuit would be likely to fail—due to the greater overload power—and a complete failure would result.

Most colour sets use a double-wound mains transformer input circuit, the primary of which acts as an autotransformer for h.t., while secondaries feed the picture tube heater in isolation and possibly a separate rectifier system for the transistorised sections of the set. Hybrid sets, too, generally have the heaters of the valves connected in series and these are fed through a thermistor from a tapping on the primary of the mains transformer. The primary winding is commonly connected to the chassis line (except in the case of all-transistor models) to provide the negative return for the h.t. circuit, so in spite of the employment of an input transformer, the chassis and metal parts of the set will be in contact with one side of the mains supply (the neutral side when the set is correctly installed).

The mains circuit also often feeds the auto-degaussing system, as explained in Chapter 6, and sometimes a microswitch on the tuner for energising the system switch solenoid.

Each circuit feed, including the mains input, is fused. The mains fuse on hybrid models is rated at about 5 A, while the main h.t. supply fuse is about 2 A and the supply feed to the transistorised sections is about 1 A.

It will thus be seen that the mains input and h.t. and l.t. supply circuits are somewhat more involved than those in monochrome sets. Nevertheless, the initial tests and observations as made to monochrome models apply equally to colour models as a means of speedily identifying the fault area. For example, lack of valve heater glow and no h.t. voltage would certainly point to trouble in the primary mains input, including the on/off switch and fuse. If there was h.t. voltage it would point to trouble in the heater circuit itself—including the heaters of the series-connected valves (excluding the picture tube which, in colour sets, is generally fed from a separate winding on the mains transformer), the mains dropper if used and, of course, the thermistor. As already mentioned, though, the use of a mains dropper is avoided in most colour sets by tapping the heater line from the primary of the mains transformer. This minimises electrical power dissipation inside the set and keeps the internal ambient temperature reasonably low, which is desirable so far as the transistorised sections are concerned.

SOUND CHANNEL SYMPTOMS

Sound channel faults are diagnosed in colour sets in exactly the same way as they are in monochrome models. There is virtually no difference in circuitry here apart from the probable use of transistors instead of valves; but even so, many monochrome sets are now changing over to transistors in the tuner, i.f. and intercarrier sound channels at least, and the trend appears to be following through to the audio sections themselves.

Dual-standard colour sets use the ordinary sound i.f. take-off on the 405-line standard and the intercarrier sound technique on the 625-line standard, while the 625-line only models use intercarrier sound exclusively, of course.

Sound symptoms in colour sets have almost everything in common with those in monochrome sets, including complete failure (i.f., intercarrier or audio areas), intermittency, vision-on-sound, intercarrier buzz and so forth.

MONOCHROME DISPLAY FAULTS

A correctly adjusted and working colour set will display good monochrome pictures from a colour transmission with the colour turned off or the saturation control turned right down, and also from a monochrome transmission, due to the lack of colour information on the transmission automatically switching off the chroma sections of the set ('colour killing'). Monochrome reproduction on a colour set results because of the inbuilt factors of compatibility—on the transmission and in the set itself.

Thus, if a colour set fails to give reasonable monochrome pictures then it is unlikely to give good colour pictures! This is a worthwhile axiom to bear in mind. Indeed, a great deal of information as to the colour performance can be obtained by studying the black-and-white display.

The most common symptom in this category is spurious colour on a black-and-white picture. Normally, there should be no evidence of a predominant colour in the monochrome tones, and the display of unwanted hues can be caused by faults or maladjustments in the purity convergence, grey-scale tracking and colour circuits themselves.

Chroma Channel Not Muted

If the colour killer and/or associated chroma control circuits are failing to work properly, for instance, the chroma section could well be open for signal even when there is no colour information on the signal. Although there is no actual chroma signal to activate the colour circuits under this condition, noise components within the subcarrier response can be interpreted by the chroma circuits as colour signals. Thus, apart from the cathodes of the picture tube receiving luminance signals (monochrome signals in this respect), the three grids will also have the noise signals passed to them from the colour-difference amplifiers, a fault that can give rise to grainy 'rainbow' symptoms on the black-and-white picture. Any symptom of coloured grain on the black-and-white picture is thus a fairly sure indication that the trouble lies somewhere in the colour killer section of the chroma channel.

It will be recalled that the chroma channel is designed to pass signals *only* when a rectified burst or burst-derived ripple signal is fed to the controlled stage. Normally—and that is on monochrome where there is no burst signal—the chroma channel is muted by back biasing of some kind. If the chroma channel is otherwise open, then something must have happened to have removed the muting bias. Remember, the muting is normally *removed* by the colour infor-

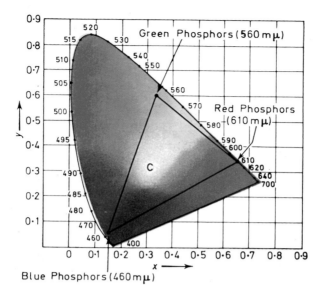

PLATE 1. *Chromaticity diagram. The triangle within the coloured horseshoe represents the range of colours obtainable from the red, green and blue phosphors. Colour wavelength is shown round the locus of the colour area.*

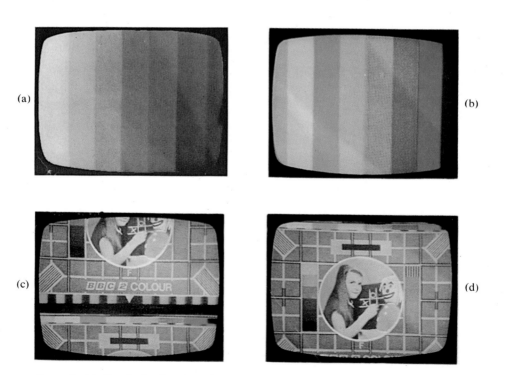

PLATE 2. *(a) Standard colour bars displayed in monochrome. (b) Standard colour bars displayed in colour. Standard colours at the top of Test Card F shown by slipping the field (c) and by vertical shift (d).*

PLATE 3. *Enlarged view of illuminated dots on the tube face as might be seen through the Beam Landing Pattern Scope on a picture (see Chapter 11 and Fig 11.6).*

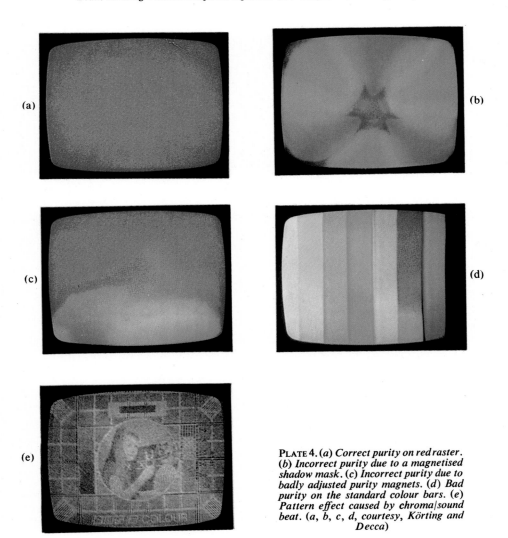

(a)

(b)

(c)

(d)

(e)

PLATE 4. (*a*) *Correct purity on red raster.* (*b*) *Incorrect purity due to a magnetised shadow mask.* (*c*) *Incorrect purity due to badly adjusted purity magnets.* (*d*) *Bad purity on the standard colour bars.* (*e*) *Pattern effect caused by chroma/sound beat.* (*a, b, c, d, courtesy, Körting and Decca*)

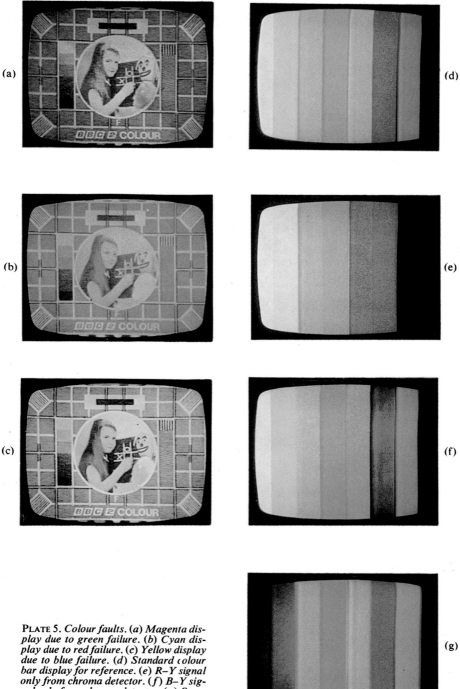

PLATE 5. *Colour faults.* (*a*) *Magenta display due to green failure.* (*b*) *Cyan display due to red failure.* (*c*) *Yellow display due to blue failure.* (*d*) *Standard colour bar display for reference.* (*e*) *R–Y signal only from chroma detector.* (*f*) *B–Y signal only from chroma detector.* (*g*) *Standard colour bars with no Y signal.* (*d, e, f, g, courtesy, Körting and Decca*).

PLATE 6. *Test Card F (see page 226). (courtesy, BBC)*

(a)

(b)

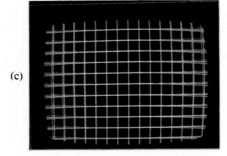

(c)

PLATE 7. *Symptoms of misconvergence on Test Card F (a) and (b). For accurate convergence adjustment a crosshatch generator is desirable. At (c) is shown the nature of the pattern produced by this type of instrument (courtesy, Pye-Unicam Ltd.). This display shows correct convergence at the middle of the screen (static convergence) but poor convergence at the edges which is correctable by the dynamic convergence controls (see page 287 and Chapter 6). However, as shown by (a) and (b) severe misconvergence is visible on Test Card F*

mation—the rectified bursts or ripple. Some sets incorporate a colour killer threshold preset which, if incorrectly adjusted, can result in the chroma channel being switched on by noise.

Chroma Hum

Although the chroma channel is muted during the reception of a monochrome signal, the colour-difference amplifiers and the clamp triodes or transistors on the three grids of the picture tube remain active. It will be recalled that the 'clamps' are actuated by line pulses and that they give the d.c. reference level to the grids relative to the cathodes. Clearly, then, *any spurious signal* that may be injected into the chroma or colour-difference channels or clamp circuits when the set is working from a monochrome transmission will cause some sort of colour to appear on the black-and-white picture. The 'colour section' vulnerable lies between the colour killer-controlled chroma amplifier stage and the picture tube control grids.

Hum in the R − Y channel, for instance, will produce a typical *horizontal* hum which is shaded towards cyan along its middle and towards red at each side; in the B − Y channel the middle of the bar will be very dark blue—almost black, depending on the magnitude of the hum—and the edges will shade towards a lighter blue; and similarly, in the G − Y channel the middle of the bar will be shaded towards magenta and the edges towards green. Inter-electrode leakage and grid-emission effects in the triode-pentodes of the colour-difference channels can precipitate these symptoms.

Hum originating in the red and blue detectors gives a slightly different symptom, depending on how it is actually produced and the hum bar—although still falling horizontally across the picture—is somewhat less dramatically shaded. Red detector circuit hum gives a predominance of red in the middle of the bar, shading towards cyan along the bottom edge and towards magenta along the top edge, while blue detector circuit hum gives the top of the picture a bluish hue, shading towards magenta at the bottom.

Horizontal shading can also arise from distortion in the line pulses actuating the clamps on the tube grids. If these pulses are delayed or advanced, due to a change in the value of of associated resistor or capacitor, colour shading effects can appear towards the sides of the picture.

Poor E.H.T. Voltage Regulation

Poor e.h.t. voltage regulation means that the e.h.t. voltage applied to the final anodes of the picture tube fails to remain constant at all levels of beam current. In black-and-white sets the effect of this shortcoming is well known—the picture becomes larger and larger as the brightness control is turned up and eventually disappears when the e.h.t. voltage drops to a low value. This is sometimes referred to as 'blooming', and a similar symptom can occur in colour sets when the e.h.t. regulator or associated circuit runs into trouble.

On colour sets, however, the symptom is more distracting, for the enlarging effective scan spot—made up of the *three* beams—tends to encourage lack of registration. The result of this is similar to misconvergence and shows as colour fringing around the white parts of the picture. A fault in the e.h.t. trebler module can be responsible for this.

An open-circuit e.h.t. regulator valve or series resistor can cause the e.h.t. voltage to rise to an abnormally high level at low beam currents, and to fall in level as the beam current increases. The abnormally high e.h.t. voltage can be dangerous from the x-ray point of view, and it will almost certainly show its presence in the form of corona discharge—with the accompanying ozone smell —or excessive arcing when the brightness control is turned right down.

The opposite fault—low e.h.t. voltage—also accompanied by poor regulation, is another pointer towards the e.h.t. regulator or e.h.t. generator; the former is caused by electrical leakage in the regulator valve or associated component and the latter to transformer, e.h.t. trebler module or valve trouble.

It is also worth noting that maladjustment of the grey-scale tracking presets, resulting in an abnormally high first anode potential on the picture tube guns, can incite an excessive initial beam current, thereby pulling the e.h.t. regulator away from its normal operating range. In this event, though, the symptoms of incorrect grey-scale tracking are also generally evident.

Poor Focus

It is unlikely that the focus of one gun will deteriorate while those of the other guns remain normal, except possibly in cases of a defect in the tube affecting one gun only. This is because the focus electrode is common to all beams. However, bad focus points to trouble in the focus electrode potential generator which, as we have seen (Chapter 7), often takes the form of an e.h.t. rectifier energised from a tapping on the line output transformer. When the e.h.t. is tapped down to give the focus potential, one of the high value resistors could be faulty.

Abnormally High Brightness

As with monochrome sets, this symptom can arise from trouble in the brightness control and video drive circuits, but is complicated in colour sets because of the nature of the brightness control circuit. This circuit generally controls the standing current in the Y amplifier and hence the potential at the tube cathodes, via the grey-scale tracking presets. It is also complicated because the three control grids are themselves fed from the colour-difference amplifiers and their associated clamping circuits.

This means, therefore, that the inability of the brightness control to turn off the picture completely points to a loss of tube bias, originating either in the Y (luminance) amplifier, colour-difference amplifiers or the grid clamps. However, if the grey-scale tracking holds over the range of the brightness control, even though the control fails to extinguish the raster at minimum setting, the fault is almost certainly common to all three guns, and will most likely be in the Y amplifier.

On the other hand, if the grey-scale tracking is badly disturbed, and one raster appears more predominant at low brightness than the other two, the trouble will almost certainly exist in the colour-difference amplifier or clamp associated with the colour that predominates. Of course, it is possible, though not all that likely, that two colour-difference amplifiers will have the same fault simultaneously, in which case the colour remaining at low brightness control setting will be that corresponding to a mix of the two associated colours.

Luminance Delay Line Symptoms

We have seen (Chapter 8) that this delay line serves to slow down the wider bandwidth luminance signals so that they arrive at the tube at the same time as the colour-difference signals. On colour, a bad or incorrectly adjusted luminance delay line will cause the colour elements of the picture to be displaced horizontally from the luminance (black-and-white) elements. On monochrome, however, an open-circuit line will disconnect the Y signal from the tube cathodes to leave an unmodulated raster. In practice, though, it is possible for such a fault to disturb the biasing of the luminance channel and so alter the tube bias as regulated by the brightness control, thereby disturbing the range of brightness control adjustment over the unmodulated raster.

A fault in the matching of the line to the input and output circuits, due to a change in the value of a matching resistor, capacitor or inductor, depending on the precise design of the circuits, can incite signal reflections to and fro along the line, resulting in the symptom of 'ringing' on the monochrome picture. Sometimes there is only one major overshoot or 'ring' from this cause, which produces a single 'ghost' image slightly displaced horizontally from the main image.

Bad Purity

Chapter 6 provides all the information required to establish the optimum purity of a picture tube. Maladjustment of the purity magnets is a common cause of colour on a black-and-white picture. To check this, extinguish all beams but the red one, when the correct purity is then seen as a perfectly red raster, as shown in Plate 4(a).

Incorrect purity caused by a magnetised shadowmask in the picture tube is shown at (b), due to incorrectly adjusted purity magnets on a red raster (c) and on the standard colour bars (d).

Misconvergence

Another common cause of colour on a black-and-white picture is misconvergence, both static and dynamic. When a set is accurately converged the colour picture tube should behave just as an ordinary monochrome one on monochrome signals. Misconvergence shows up as displaced primary colour components, in the middle of the screen with static misconvergence and towards the edges with dynamic misconvergence. The most common symptom is that of dynamic misconvergence, and truly perfect dynamic convergence is rarely attained in practice with a domestic receiver, but the residual misconvergence should not be so severe as to show as colour at the edges of a monochrome picture at normal viewing distance. Close scrutiny, however, almost always shows very slight colour contamination on all but high quality and expensive monitor receivers.

Bad Grey-Scale Tracking

Again, this is another common cause of colour in a black-and-white picture. Unless the glowing intensities of the red, green and blue phosphors hold correctly for a white display over the *full range* of (*a*) the brightness control and (*b*) the luminance signal swing, colour tinting will affect the monochrome picture.

Chapter 8 contains all the information required to secure the best possible grey-scale tracking performance from most sets. It is not intended to repeat this here, but both conditions (*a*) and (*b*) above can be influenced by incorrectly adjusted tube first-anode and cathode drive presets. A disconcerting effect is the progressive change in overall hue of a monochrome picture with changes in brightness control, which is made even worse by changes of different hue in parts of a picture as actually governed by their brightness. For instance, a medium grey may show as grey, while a very light grey or white might go, say, towards yellow, and a dark grey might swing more towards a blue shade. This sort of symptom should lead immediately to a check of grey-scale tracking and to those parts of the circuit which are related *dynamically* to the beam currents of the three guns.

Picture Size, Linearity and Centring

Prior to making any adjustments to the convergence, it is essential to ensure that the picture is correctly centred in the screen, and that it is of correct aspect ratio and linearity (see Chapter 11). Should a fault develop in the timebase circuits that affects any of these factors, then a previously correct convergence performance is likely to be impaired. The trouble can be complicated by attempting to mask the real fault by readjusting the dynamic convergence presets.

Faulty Synchronisation

Line and field synchronising faults produce symptoms which are the same as those on monochrome-only sets, but colour sets almost always adopt a flywheel-controlled line sync circuit, as distinct from direct line sync. This sort of circuit can get into trouble over its phasing as well as its locking, the former causing the picture to slip sideways *within* its raster, while remaining in lock, and the latter causing the picture either to fail to lock horizontally at all (giving the effect of 'horizontal rolling') or to fall out of lock intermittently, possibly when changing channel. Symptoms of this kind and their remedies are fully covered in the *Television Servicing Handbook*.

The symptoms associated with field sync defects are very well known, such as picture judder, vertical rolling, weak vertical lock and so forth, and are also covered in the above mentioned book.

For the best colour pictures the interlacing must be very accurately handled by the set, and designers take a great deal of care to attain this by the use of an interlace filter circuit between the sync separator and the field generator, by efficient screening in and around the line timebase section, and by well designed filtering and decoupling in the field generator circuits.

No Raster

We have already seen that this symptom can be caused by (*a*) lack of or wrong electrode potentials, (*b*) incorrect tube biasing and, of course, (*c*) lack of e.h.t. voltage. Some models feature the basic e.h.t. system, whereby an overwind on the line output transformer generates pulses of sufficient amplitude which, after rectification, correspond to an e.h.t. potential of around 25 kV (or about 8·3 kV when an e.h.t. tripler module is used—as in recent single-standard sets), while others use a separate e.h.t. generator (often with a voltage multiplier) which itself is activated by line pulses. Most models, therefore, rely in some way on the line timebase for their e.h.t. voltage, so failure of this section of the set means lack of e.h.t. and no raster.

The line whistle on 405 lines (dual-standard models) gives a good clue as to whether or not the line generator is working, but very few people can hear the 15,625 Hz whistle on the 625-line standard.

It is certainly not a good idea to test the e.h.t. pulse potential or d.c. by drawing an arc from a 'live' point on the line timebase or e.h.t. circuits. While this method of e.h.t. checking is practised extensively in the servicing of monochrome sets, it can give rise to several problems in colour sets. Most of these are now hybrid designs, and large amplitude discharges of the kind mentioned can destroy or, at least, change the characteristics of the transistors in the set. Moreover, colour set e.h.t. circuits are capable of delivering 30 W or more at 25 kV, which is virtually lethal power; but, in addition, picture tube and valve electrons accelerated by this potential incite x-ray emission on impact, which can be harmful to a person working for long periods very close to an unshielded e.h.t. section.

Raster too Small

A smaller than normal raster in good focus is an indication that the e.h.t. voltage is excessive, and steps should be taken to check this and make the necessary adjustments.

Raster too Large

A larger than normal raster in poor focus is an indication of low e.h.t. voltage.

WHAT THE CONTROLS CAN TELL

As in monochrome set servicing, the effect that the various controls have on the display can give the technician clues as to the exact whereabouts of a fault. As a simple illustration, if the field is just about locking with the vertical hold control set hard to one end of its range, then this points definitely to trouble in the time-constant ('timing') components of the field generator—possibly a change in the value of a resistor connected in series with the vertical hold control would be responsible.

Clues of this kind, which have been discovered during the course of servicing monochrome sets, can also apply to black-and-white fault symptoms in colour sets. However, colour sets have, at least, one extra main control—the saturation control—and hosts of extra presets! The behaviour of the fine tuning control (u.h.f. tuning) can also affect the performance on colour much more than on monochrome, and it is just as well to see why this is so because its effect on the display can help to locate the area in which a fault exists.

Tuning Control

When the tuning control is turned the local oscillator frequency is altered which, of course, alters the intermediate frequencies. Fig. 12.1 gives an idea of the overall i.f. response curve of a colour set and the positions of the 'carriers' (at i.f.) along it. This shows the vision carrier on the far side of the curve at about 6 dB down and the sound carrier filtered down to a much lower level for correct intercarrier sound operation. The colour subcarrier (about 4 dB down) occurs at a point 4·43 MHz (approximately) from the vision carrier, which corresponds to the colour subcarrier frequency, while the sound carrier occurs at a point 6 MHz from the vision carrier, which corresponds to the intercarrier frequency.

Now, these signal relationships are not affected at all by adjustment to the u.h.f. tuning (within the bounds of a channel, of course!) This is the reason that 625-line standard sound is virtually insensitive to fine tuning errors—the difference between the sound and vision signals always remaining at 6 MHz. The same also applies to the colour subcarrier. This always remains at 4·43 MHz, irrespective of the setting of the fine tuning within the bounds of a channel.

The correct tuning point on 625-line standard monochrome is obtained simply by adjusting the control for the best picture definition, consistent with minimum intercarrier buzz. With colour, however, this method of tuning is insufficiently accurate because the colour subcarrier is the most critical signal and has to be placed accurately on the response curve for the best colour performance.

The method used, therefore, is first to adjust as for monochrome, and then to readjust very slightly for the best colour display (using Test Card F, for instance),

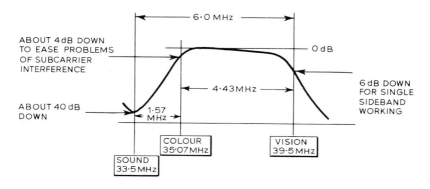

FIG. 12.1. *Overall vision response curve, showing the placement of the carriers.*

consistent with the smallest amount of pattern interference. When this is achieved the ratios of signal levels along the response curve are automatically optimised for the best results, assuming that the i.f. channel is correctly aligned.

Sets with a.f.c. permanently in circuit will automatically pull-in to the correct tuning point, but some have switched a.f.c., it being switched on after establishing the correct tuning point.

When the tuning control is adjusted the response curve is unaffected, it remains exactly as created by the i.f. alignment, but the three signals move along the response together in a direction depending on which way the tuning is adjusted. It is possible, therefore, to mistune to the extent of losing the colour subcarrier altogether, the effect being that the subcarrier slides down the side of the response. This might also result in a loss of sound, but excessive undulation at the top of the response curve, carried by bad i.f. channel alignment, can affect the colour signal while keeping the sound carrier at a fairly reasonable level due, perhaps, to a subsidiary response a little outside the normal passband.

Conversely, the subcarrier amplitude (at the vision detector) can be altered by the vision carrier slipping too far down the far side of the response curve. This can also affect the colour performance, causing a progressive reduction in the saturation and finally a complete loss of colour.

The correct tuning point is not particularly difficult to establish because of the effect that the various rejectors in the i.f. channel have on the picture (see Chapter 15). The sound i.f. signal in the i.f. channel, for instance, is correctly attenuated by the sound rejector only when the fine tuning is accurately adjusted because then the signal frequency coincides with the rejector frequency. Incorrect tuning causes the sound i.f. signal to rise from the dip in the response produced by the rejector and this results in intercarrier buzz.

Moreover, too high a level of sound i.f. signal in the i.f. channel can produce a beat signal between the sound and colour subcarrier signals of an abnormally high amplitude that cannot be attenuated sufficiently by the sound rejector. This can cause severe beat patterns on the picture when colour information is present on the received signal (see Plate 4(e)).

To sum up, therefore, on one side of the correct tuning point, the colour subcarrier falls too far down the response curve and results in desaturated (or nonexistent) colours, while on the opposite side the colours tend towards oversaturation, and beat patterns appear on the picture. The correct tuning point can be found by adjusting for the highest colour saturation, consistent with maximum luminance definition and minimum intercarrier buzz and beat patterns.

If the effect of tuning differs substantially from the foregoing description it might be as well to check the overall alignment of the set.

Colour Saturation Control

This control simply regulates the level of the chroma signal applied to the colour detectors from the chroma amplifier. Lack of operation of this control would indicate trouble in the chroma amplifier section, but it should be remembered that as the contrast in monochrome sets is subjected to an automatic control by the vision a.g.c. system, so the saturation in colour sets is subjected to automatic control by the a.g.c. system. It will be recalled that a d.c. derived from the bursts is used as the control bias.

Tint Control

This is an arrangement whereby a differential adjustment is provided for the biasing of the three guns of the picture tube. The nature of 'white' can thus be modified for personal preference. This control has no parallel with the tint control of NTSC sets, which actually changes the hue by regulating the phasing of the signals in the chroma circuits.

Incorrect operation of the tint control would point to trouble in the tube biasing circuits and/or in the grey-scale tracking.

Convergence Controls

Failure of the convergence presets to influence the three-colour displays in the manner expounded in the maker's service manual would certainly indicate trouble in the convergence circuits or in the convergence yoke itself. In cases of doubt, however, it is best to run right through the recommended convergence procedure—both static and dynamic. Most of the other controls of a colour set work in a manner similar to those of monochrome sets.

As we have seen, many colour faults are highlighted when the colour set is working from a monochrome transmission. There are, of course, hosts of other fault symptoms that appear only when the set is reproducing colour from a colour transmission, and we shall now see how these can be located to particular areas of the chroma stages.

No Colour, Monochrome Only

This is the most obvious symptom in colour set servicing which indicates clearly that the trouble lies somewhere between the vision or chroma detector and the colour-difference amplifiers. If the resulting monochrome representation of the colour picture is accurately displayed without colour contamination, then one can be sure that the picture tube grids are well clamped to black level and that the colour-difference amplifiers are not introducing spurious, interfering signals. The fault would most likely lie in the chroma channel, colour detectors or reference generator. Lack of spurious colour might well indicate a muted chroma amplifier caused by failure of the reference generator or phase detector when the colour killer is designed to be biased-on by the *ripple signal*. If the set has facilities for disabling the colour killer (e.g. a threshold preset), then this should be activated, to enable the chroma channel to pass signal. If correct colour is still not obtained and the effect is that of noise or, indeed, lack of colour sync, then the real trouble would be located in the chroma amplifier, or the reference generator section (incorrect frequency can mute the ripple signal), or the colour sync (burst gating, phase detector and reference generator control) circuits. Chapter 13 describes the servicing procedures that can be applied to a suspect section once it has been located.

Since the monochrome picture is correctly reproduced, we can be sure that the parts of the set up to the video detector and thence to the picture tube cathodes, via the luminance amplifier sections, are in perfect order. However, if the monochrome definition is very poor, something could be amiss in the i.f. stages—or even in the tuner or aerial—to cause discrimination against the band of frequencies carrying the colour information sidebands. A tuner failing to tune to the channel correctly is one possibility, and in cases of colour failure it is always as well to check the tuning, especially when the monochrome definition is poor. A mass of metal in proximity to a u.h.f. serial can even put a dip in the response of a colour channel at a frequency corresponding to the colour subcarrier, and thus absorb most of the colour information from the signal, leaving the display in black-and-white only! An aerial of the wrong channel group can also roll-off the chroma. Similar troubles can also occur through a badly aligned, designed and matched aerial amplifier or v.h.f. relay system.

If the set features a colour-killer threshold control, this can be advanced to activate the chroma amplifier (or, alternatively, arrangements can be made to remove the muting bias from the controlled chroma amplifier stage), and if this results in coloured grain on the screen when the aerial is disconnected, then one can be fairly sure that the chroma stages, at least, are working to some extent, for coloured grain will only occur when the chroma stages pass signals to the tube grids, via the colour-difference amplifiers. The wideband noise signals generated by the tuner are, in fact, processed by the chroma circuits within the spectrum normally embraced by the real colour information—e.g. about 1 MHz each side of 4·43 MHz.

Desaturated Colour

This symptom, at least, indicates that the sections of the set synchronising the colour information are working properly. The trouble is caused by low amplitude signals in the chroma channel and, as with the previous symptom, a drop in the response at the colour subcarrier frequency could be responsible. An alignment check might, therefore, be worthwhile if the monochrome (luminance) definition is poor and one can be sure that the aerial and/or associated equipment—amplifiers, filters and so forth—are in good order and not acting in a frequency-discriminating manner.

In the PAL system, excessive phase distortion of the composite signal—due to propagation and/or aerial shortcomings or maladjustment of the signal-carrying circuits of the set—can reduce the saturation quite a bit if the phase error is large. This, it will be recalled, happens instead of the colours of the picture changing—a by-product effect of the PAL system.

On some sets with automatic frequency correction (a.f.c.), desaturation effects can arise from incorrect adjustment of the control circuits themselves. For example, if the control voltage tends to pull the set off tune through i.f. misalignment or misalignment of the a.f.c. discriminator, then the reproduction will most certainly have weak colours. It is very important, therefore, to adjust the a.f.c. circuits exactly as detailed by the maker.

A typical symptom of faulty a.f.c. is weak colours accompanied by very critical tuning, or the disappearance of the colour on reselecting a channel carrying a colour transmission.

No Colour Synchronisation

The colour is effectively synchronised by the bursts which phase-lock the reference generator. If the reference generator remains active, yet fails to phase-lock to the bursts, the symptom is rather similar to that of line sync failure but on the colours only. The colours break up across the picture into horizontal bands and drift in a random manner if the reference generator is completely devoid of control potential. This sort of symptom points to a defect somewhere between the phase detector (discriminator) and the reference generator, via the d.c. control amplifier if used. It could also be caused, of course, by failure somewhere in the channel carrying the bursts from the gate to the phase detector. It is noteworthy, however, that incorrect reference generator 'locking' can sometimes mute the phase detector ripple signal, thereby presenting the colour killer stages in the chroma amplifier from being biased on.

However, if the colour in the picture is 'locked', yet not properly synchronised and of the wrong hue, the trouble could be caused by the generator running at an incorrect frequency. These two conditions have an analogy in the line timebase, the first is line lock failure because of a total lack of sync control on the generator which gives random line drift, and the second is due to the timebase running at an incorrect speed, but with the sync control present on the generator. The analogy is somewhat more accurate when considered relative to a flywheel-controlled line timebase. However, note the remarks earlier made about the effect of wrong-frequency reference generator signal on the ripple signal.

Incorrect Colour Displays

The most obvious of incorrect colour displays arises from failure in the chroma detectors, in the colour-difference amplifiers or in the primary colour circuits including the tube guns. For example, failure of the red gun or associated circuit will produce a display devoid of red regardless of the setting of the colour control; similarly with the colours corresponding to failure of the green and blue primary colour channels. Symptoms resulting from faults of this kind are shown in Plate 5, where (a) is a magenta display due to green failure, (b) a cyan display due to red failure and (c) a yellow display due to blue failure.

However, a fault in the actual colour-difference channel can give different symptoms, so we must be careful when considering fault displays arising from troubles in these circuits. Remember that a fault occurring between the chroma detectors and the colour-difference circuits will affect all three drives, not just $R - Y$ and $B - Y$, owing to the $G - Y$ matrix.

Further, a missing colour-difference drive will not necessarily completely delete the associated hue from the display. This is because the signal is a *colour-difference* one, meaning that it is possible for a gun not receiving drive at its grid to yield a beam from the Y signal at its cathode. This applies to sets where the primary colour matrixing occurs in the tube guns.

Thus if there is no colour-difference signal at the grid there will be no Y signal to cancel it at the cathode. The gun would then be biased by the Y signal level alone, causing a beam current. It is best to check the drives with an oscilloscope at the grids, for colours can be affected by the setting of the main and preset controls and by the saturation of the displayed signal.

Standard colour bars are left-to-right white, yellow, cyan, green, magenta, red, blue and black. Faulty $B - Y$ drive might change them to white, off-white, dull-green, bright-green, red, dark-red, very dark blue or black and black. Trouble in the G − Y drive could yield white, yellow, mid-blue, deep green, pale magenta, orange-towards-red, blue and black. Faulty R − Y drive could give white, yellow, pale-mauve, dark-green, blue, dark-towards-red, blue-towards magenta and black. In other words, faulty R − Y affects the cyan, green magenta and red bars, faulty G − Y the cyan, green, magenta and red bars again but in a slightly different way, while faulty B − Y affects the yellow, cyan, magenta and blue bars. The extent of the colour-difference fault will produce subtle variations in the incorrect hues. Since the G − Y signal comes from the matrix, assuming that the R − Y and B − Y channels are active, then a G − Y fault symptom can only be caused by matrix or G − Y amplifier trouble.

Failure of one of the colour-difference signals at the output of the chroma detector can give a *slightly* different set of symptoms, as shown in Plate 5. At (d) is a display of the standard colour bars as a reference, at (e) the display with the B − Y signal missing and at (f) the display with the R − Y signal missing. Notice at (e) the lack of blue and yellow and the change in the colours of the other bars, with red predominant, and at (f) the lack of red and the blue predominant.

Symptom (f) can be caused by trouble in the red detector PAL switching diode. A short- or open-circuit here could reduce or mute the red output and the only input into the colour-difference stages would then be the B − Y signal.

It will be appreciated, of course, that the green colour-difference amplifier will not be devoid of signal should either the B − Y or R − Y input fail, because the matrix will convey some degree of the active signal to the G − Y amplifier, which is why the colours of the bars change in the manner shown at (e) and (f). It is also the reason that the symptoms differ slightly should failure of one of the colour-difference amplifiers, as distinct from the input R − Y and B − Y signals, delete one of the colour-difference signals actually fed to the picture tube grid. For example, failure of the R − Y channel *after* the matrix would not affect the G − Y input to the tube.

Sets with primary colour signal matrixing outside the tube can produce different symptoms, but the principles expounded remain the same.

Failure of the Luminance Signal

If the luminance signal fails in the luminance amplifier section, the monochrome components and the components of detail disappear from the picture. The symptom is then that of colour in large blocks with hardly any definition at all. The effect on the standard colour bars is shown at (g) in Plate 5, from which it will be seen that white is missing and that the bars close to white are changed in hue.

Failure of the signal actually in the luminance section might not affect the stage biasing, and the brightness control will continue working normally. However, if the signal failure is actually caused by a fault in the luminance amplifier itself, then it is likely that the biasing of the picture tube will be affected, with failure of the brightness control action. Indeed, it is possible that the picture tube will be back-biased, in which case the luminance-less display might not then be seen.

Beam current limiting current faults can also offset the luminance and brightness of the display, and sometimes the colours.

Timebase Symptoms

As already mentioned, many of the symptoms resulting from trouble in the timebase circuits will be similar to those experienced in monochrome sets. However, there are several complications in colour sets due to the action of the dynamic convergence; and sometimes picture tube protective circuits are incorporated which come into action when the scan signal fails. Moreover, the field timebase is usually energised from the boost voltage, so that if this drops the field timebase ceases to work correctly. Faults can, in fact, develop which back-bias the line output valve, thereby killing not only the normal operation of the line timebase (making the line output transformer a suspect) but also the field timebase due to the drop in boost voltage.

This sort of fault, especially when it is associated with an inbuilt protection or safety circuit, as in the Philips G6 colour chassis, should be borne in mind during the course of servicing.

The timebases also nearly always activate a pincushion correction 'transductor', and this can develop a fault and reflect timebase-type symptoms. It has been known for a short in this component to over heat a resistor in the field timebase and cause it to go open-circuit, with a consequent failure of field timebase.

Signals for the dynamic convergence coils are derived from the timebases, as is now well known, and since it is necessary to 'tailor' these signals very carefully with the dynamic convergence presets to secure the exact correction of the three beams, it follows that any change in the performance of the timebases will be reflected as misconvergence. Thus, a fault gradually developing in one of the timebases will not only show as a timebase fault symptom (e.g. cramping at the bottom, top or sides, alteration in the height, width, linearity and so forth) but also as some degree of misconvergence, depending on the nature of the circuit and the timebase in trouble.

In such cases, one must never be tempted to clear the trouble by readjusting the convergence presets, for accurate convergence under these conditions is unlikely to be achieved. First, of course, the real timebase fault must be cleared, after which the set should be reconverged. Indeed, lack of convergence, or difficulty in obtaining a good convergence performance, could point to trouble in the timebases.

This is why it is essential to adjust accurately the height, width (or the correct aspect ratio on Test Card F), linearity, centring and picture levelling controls *before* adjusting the convergence. For similar reasons, it is also essential to ensure that the set is adjusted as accurately as possible to the applied mains input voltage. It is also a very good idea to let the set run for ten minutes or so, at least, before making any adjustments of a critical nature. The customer, too, should be informed that it will take at least ten minutes for the set to settle down into accurate adjustment each time after switching on.

Convergence Symptoms

If the timebases produce correctly proportioned rasters yet a fair degree of misconvergence cannot be corrected by the convergence presets, adopting the exact procedure as laid down by the maker, it is then likely that trouble of some kind exists in the convergence circuits themselves. It is not uncommon to find an open-circuit convergence coil or a break in an associated connecting wire or printed-circuit board conductor, so it is as well to check such obvious things first of all.

The next step is to work carefully through the convergence procedure, for defective circuits are pin-pointed when adjustment of the associated preset fails to give the expected results.

Under-convergence means that the convergence fields are insufficiently strong to get the three beams to strike the adjacent phosphors corresponding to the three colours, in which case the beams run approximately parallel to each other, causing a small picture element to be composed of displaced red, green and blue lights, instead of compounded colours, giving, say, white light.

Over-convergence occurs when the convergence fields are greater than required for optimum correction, a fault that could point to maladjustment or change in value of a component in the convergence circuits. Under- or over-convergence is not too difficult to diagnose when one is fully accustomed to the effect that the convergence presets have on the display. Some sets react slightly differently from others in this respect, so an intimate knowledge of the set is necessary.

Dual-standard sets have virtually doubled-up dynamic convergence presets, as past chapters have shown, which although an added complication, can sometimes facilitate the diagnosis of a convergence defect. For example, if the set gives a good convergence performance on, say, 405 lines, but not on 625 lines, the preset circuit of the latter standard, which fails to give the same adjusting effect as its partner on the former standard, should be carefully investigated, for a change in value of a component or a defunct standard-change switch contact could well be responsible.

It will be recalled that diodes are used in the line (and sometimes field) convergence circuits to give a d.c. reference so far as the line convergence waveform is concerned. Dual-standard sets use two such diodes (switched between standards) with associated series resistors—sometimes preset—so as to balance the residual d.c. in the line convergence coils over the two standards. These diodes effectively 'clamp' the convergence waveform to a fixed potential, or to effectively zero potential at the midpoint of the waveform. This action retains accurate static convergence at the centre of the screen.

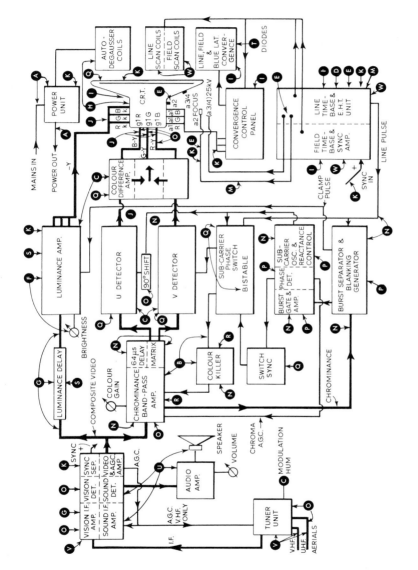

FIG. 12.2. *Example block diagram of colour set (RBM) on which the ringed capital letters identify the areas in which the fault symptoms given in the text could exist.*

247

Should such a diode fail, a symptom similar to that of incorrect static convergence will result, and this will show on a dot pattern as vertical displacement of one of the colours. If the blue beam is displaced, the affected dots will appear as basically yellow with blue slightly below—the effect gradually tailing off towards the sides of the screen.

Sets with a preset resistor in series with the diode will exhibit no effect when the preset is adjusted. A defective diode on one standard will, of course, show the symptom only on that switched standard, the other standard picture being without the fault.

Clamp diodes are used only in the line convergence circuits because of the lack of d.c. reference on the activating signal waveform from the line timebase. In the field circuits, the activating signal waveform generally has a d.c. reference, which is why diodes are not always used here.

SUMMARY DIAGRAM

A fault-finding summary of this chapter is given in relation to the example block diagram in Fig. 12.2 (RBM, Models CTV25 and CV2510) and the fault symptoms listed below. The reference letters on the block diagram show the areas of the set in which the fault is likely to lie to produce the associated symptoms.

Complete Failure (A)
Chroma Channel Not Muted (B)
Chroma Hum (C)
Poor e.h.t. Voltage Regulation (D)
Poor Focus (E)
Abnormally High Brightness (F)
Luminance Ringing and 'Ghosting' Symptoms (G)
Bad Purity (H)
Misconvergence (I)
Bad Grey-Scale Tracking (J)
Faulty Synchronisation (K)
No Raster (L)
Too Small/Too Large Raster (M)
No Colour, Monochrome Only (N)
Desaturated Colour (O)
No Colour Synchronisation (P)
Incorrect Colour Displays (Q)
Spurious Colours on Monochrome (R)
No Luminance Signal, Colour Only (S)
Lack of Screen Centre Convergence (T)
No Sound (U)
No Sound or Vision, Raster Normal (V)
Defective Scans (W)

CHAPTER 13

Servicing Procedures

THERE ARE SPECIFIC procedures that can be advantageously adopted during the course of servicing colour receivers. Some of these have already been referred to in Chapters 12 and 14, but there are others which require test instruments and temporary circuit changes in order to check a suspect section, pinpoint a fault and help with readjustment.

It is not proposed to deal with pre-vision detector, sound channel and timebase faults in this chapter for, apart from misalignment troubles which can sometimes influence colour more than Y definition (considered in Chapter 15), these yield symptoms that have a great deal in common with those of monochrome sets. This chapter, therefore, is concerned essentially with the chroma and Y signals leaving the vision detector, the colour-difference signals leaving the red and blue synchronous detectors, the reference signal, the colour bursts leaving the gate, and the ripple signal leaving the phase detector.

DISABLING THE COLOUR KILLER

If the fault is normal monochrome but lack of colour then, as mentioned in Chapter 12, we might at least be able to obtain an indication as to the faulty area by disabling the colour killer. This really means biasing-on the colour-killer controlled chroma amplifier. It is designed for the biased-off condition, except when colour bursts are present on the transmission (i.e. when the transmission is colour encoded). The swinging bursts acting in the phase detector associated with the reference oscillator produce the ripple signal, which after rectification yields the potential for biasing-on the controlled chroma amplifier. Without bursts the chroma section remains inactive.

When the killer is disabled on a monochrome transmission, spurious signals (noise, for instance) focused within the chroma passband will penetrate the chroma amplifiers to the red and blue detectors and the colour-difference amplifiers and appear as coloured grain, colloquially known as 'confetti', on the screen of the picture tube. If biasing-on the chroma amplifier restores the colour on a colour transmission, one could then be sure that the trouble area embraces the *biasing-on* circuits of the killer section, including the controlled chroma amplifier itself, the control potential circuits from the ripple signal output of the phase detector, the ripple amplifier, the ripple rectifier and the d.c. amplifier to the control potential input of the chroma amplifier, if used.

If the biasing-on action has no apparent influence on the monochrome display—for instance, no confetti on a monochrome transmission and no colour on a colour-encoded transmission—then either the chroma amplifier channel is failing to pass the chroma signal or the reference generator is defunct. While it is not difficult to understand why colour noise would be suppressed by signal discontinuity in the chroma amplifier (from the video detector output to the inputs of the red and blue synchronous detectors), it is not so obvious why the same effect can sometimes result from failure of the reference signal in the set.

The reason is that the red and blue detectors are actually switched on and off by the reference signal, and if this signal is absent they may be effectively 'switched off' (depending on circuit design) and thus fail to pass the chroma noise signals to the colour-difference amplifiers, and thence to the grids of the tube.

HOW TO DISABLE THE COLOUR KILLER

Some sets have facilities for biasing-on the chroma amplifier in the absence of bursts and others require a temporary circuit change or an addition to instigate the action. Fig. 13.1 shows the basic circuit of the colour-killer amplifier and the controlled chroma amplifier pertaining to the RBM series of receivers. In the absence of control potential from the collector of TR1, TR2 is devoid of base

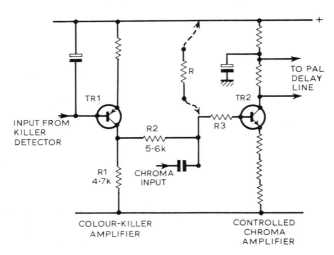

FIG. 13.1. *A method of disabling the colour killer to cause the chroma bandpass amplifier channel to pass signal independently of the killer bias.*

bias and is thus non-conducting. However, when a negative potential is applied to TR1 base from the colour-killer detector, the transistor conducts, and the negative fall in potential at its collector is communicated directly to the base of TR2 as a positive forward bias, which switches the transistor on and causes it to pass chroma signal to the PAL delay line, and thence to the colour detectors.

It should be noted, of course, that TR1 is pnp and TR2 npn, the former calling for a negative and the latter for a positive switch-on bias.

Clearly, then, it is a relative simple matter to over-ride the killer and get the controlled chroma amplifier to switch on in the absence of bursts on the transmission by applying a suitable bias from some other source. In Fig. 13.1 this is done by connecting the base circuit of TR2 to a positive potential (the positive supply rail) through a hold-off or current limiting resistor, labelled R in the circuit.

The value suggested by RBM is 18 kΩ, but this is somewhat related to the potential-divider effect created by R itself in the top arm and R1 and R2 in series in the bottom arm. Such a value, therefore, cannot be taken as suitable for all circuits. Most service manuals state a value for this resistor or suggest an alternative method for activating the chroma amplifier.

In some sets it would appear to be more convenient to over-ride the killer bias on the chroma controlled amplifier by applying a counter-bias from a small 3 V or 4·5 V battery. It is always best to apply the bias through a reasonably high value resistor, depending on the design of the controlled amplifier and on whether it uses a valve or transistor. It is essential to get the battery polarity correct to avoid damaging a valve, transistor or semiconductor diode. Where the controlled stage uses a transistor, therefore, due respect must be paid to its type, since the base counter-bias required by a pnp device is negative and by an npn device positive, as already intimated.

OSCILLOSCOPE TESTING

It is possible that the action of disabling the colour killer, in conjunction with the reasoning given above, will lead to a speedy location and correction of the fault condition. However, if the fault persists it can quickly be tracked down by using an oscilloscope to display the video waveforms in the luminance and colour stages.

One must have a colour-encoded input signal, of course, from which to derive the waveforms, and by far the best signal for this purpose is that corresponding to the transmission of the standard colour bars. These are sometimes transmitted by the authorities but are always present at the top of Test Card F—visible when the height of the picture is reduced (see Chapter 11—they are not always available just when wanted for fault-finding, so it is desirable to have a source of such signal at hand in portable form, and this is where the colour-bar generator can prove of considerable value.

The colour-bar signals are fully analysed in Chapters 10 and 11, as also is the type of instrument that can create them. Now, let us suppose that we apply a u.h.f. colour-bar signal to the aerial input of the set, and tune the set to respond to the signal. Under correct working conditions the set would reproduce the bars almost exactly as shown in Plates 2(b) and 5(d). However, the bars would be displayed in stepped-grey, monochrome, when the symptom is that of total colour failure.

By connecting an oscilloscope of suitable bandwidth (see Chapter 11) across the video detector load with its Y-input to the 'live' side of the load and the earthy side to set chassis, taking due precautions regarding mains isolation on 'live chassis' sets, a display similar to that shown in Fig. 13.2(*a*) will be obtained when the oscilloscope's X sweep is suitably synchronised or triggered.

This shows one line of composite video waveform, complete with the stepped luminance values upon which are superimposed the V and U chroma signals and the colour burst. The falling side of the line sync pulse—which triggered this oscillogram—is just visible on the left of the burst. This sort of display indicates without doubt that the complete colour-encoded signal is, at least, being demodulated properly by the video detector, and it is comforting to know that colour information is present at this point, even though the set itself is failing in its reproduction.

At this juncture it is instructive to observe the waveform at the same point with the colour switched off, which then represents the Y signal, as shown in Fig. 13.2(*b*). In this oscillogram one line sync pulse is brought into the picture on the right-hand side. Some colour-bar generators have facilities for switching on the bursts while switching off the V and U chroma signals, and Fig. 13.2(*c*) shows the waveform of this kind of signal with the oscilloscope adjusted to display one line with two line sync pulses and associated bursts. To complete this series of primary waveforms that can be obtained across the video detector load, Fig. 13.2 shows the signal encoded with the V chroma alone at (*d*) and with the U chroma alone at (*e*). Since both (*d*) and (*e*) are colour-encoded, the bursts are, of course, also present on them.

The display in Fig. 13.2(*b*) is pure monochrome and the stepped waveform results from the nature of the Y signal of the colour bars.

TESTING THE CHROMA BANDPASS AMPLIFIER

The composite signal (Fig. 13.2(*a*)) is ideal for testing the chroma bandpass amplifier, and the correct chroma signal that should be displayed at the output of this amplifier and the decoder is shown in Fig. 13.3. This represents the composite chroma envelope of one complete line of colour information, and the salient feature of the display is the symmetry of the positive and negative halves of the signal relative to the centre zero-axis. Such a display would, indeed, indicate that the chroma amplifiers are properly adjusted in relation to the vision i.f. channel alignment and that they are handling the chroma signals very successfully.

Fig. 13.4 gives the chroma bandpass amplifier circuit of the Decca CTV25 series sets, and assuming that the amplifiers are working correctly, the waveform of Fig. 13.3 should be present at the collectors of TR600 and TR601. If the signal is present at the first transistor but not at the second, then it is possible that the controlled stage (TR601) is failing to switch on due to lack of colour-killer bias, and this should lead to attention of the circuits from which the bias is derived. It has already been explained how a fault of this kind can be proved by biasing-on the controlled chroma amplifier with a battery or resistor.

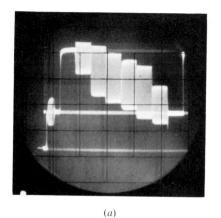

(a)

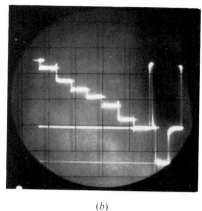

(b)

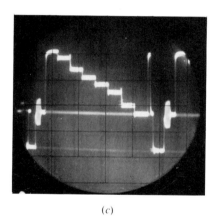

(c)

FIG. 13.2. *Oscillograms at the video detector.* (a) *composite colour-encoded signal,* (b) *Y signal with the chroma switched off,* (c) *Y signal plus bursts but no chroma,* (d) *composite signal without the U chroma signal and* (e) *composite signal without the V chroma signal.*

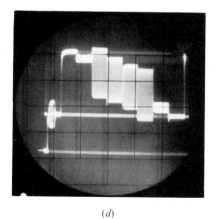

(d)

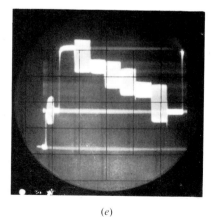

(e)

Clearly, then, once a chroma signal is applied to the bandpass amplifier, it can easily be traced right through the various stages by observing the actual signal on the screen of an oscilloscope. This is a bonus derived from the fact that the chroma is effectively a 'video' signal so far as the output of the video detector is concerned. We cannot normally adopt this method of signal tracing

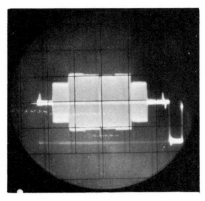

FIG. 13.3. *Subcarrier modulated by V and U chroma signals, showing also a trace of the bursts. Note: The line sync bursts are often severely attenuated or eliminated at the decoder input by line pulses applied to a chroma amplifier transistor. See, for example, the line blanking input in Fig. 13.4.*

in the vision i.f. stages, however, because the frequency of the signal is generally too high. The chroma envelopes, of course, are patterned by virtue of the 4·43 MHz subcarrier, so it is to the subcarrier that the oscilloscope is called upon to respond.

Let us now return to the chroma waveform in Fig. 13.3. This has lost its d.c. identity because of the a.c. coupling from the video detector to the input of the amplifier, so it should always appear symmetrical to the axis. Fig. 13.5 shows the waveform (two lines of chroma this time) in asymmetrical form, which could arise from a chroma stage running with an incorrect bias potential. The symptom in Fig. 13.5, in fact, would be most likely caused by a low control voltage, giving the wrong operating point.

As previously shown, the chroma signal is often stabilised by an automatic gain control system (automatic chroma control or a.c.c. for short), the control potential for which is related to the burst amplitude, and derived from the burst signal after rectification—from a similar source to that which provides the colour-killer bias.

The biasing of the stages, including that provided by the colour-killer source and the a.c.c., therefore, should normally be adjusted so that the symmetry of the chroma waveform is retained reasonably well over the full range of a.c.c., colour saturation control and working input signal levels. The input signal amplitude, of course, automatically controls the chroma gain by way of the a.c.c.—that is by way of the amplitude of the bursts. If the colour-encoded generator has a burst amplitude control, then the effect of changing the amplitude of the bursts can be seen on the displayed chroma signal. In a correctly adjusted chroma channel, the amplitude of the chroma signal should remain substantially constant over almost 20 dB of change in burst amplitude—depending on the precise design of the set—and the signal symmetry should be retained.

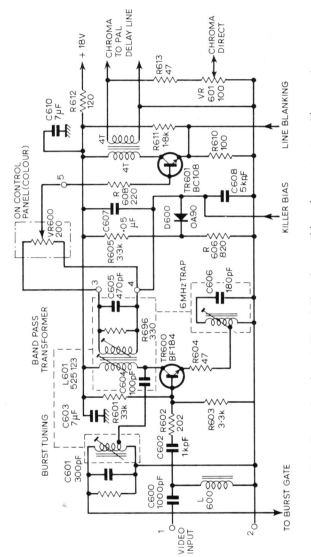

Fig. 13.4. *Chroma channel circuit (Decca). Note: Later models employ an extra stage with automatic chroma control.*

255

Considerable asymmetry should certainly lead to a check of the biasing of the chroma amplifiers, the a.c.c. system and the colour-killer bias input. An electrical leak in the coupling capacitor to the input of the chroma amplifier is another possibility that should be investigated. It will be appreciated that the ordinary vision a.g.c. system will counter the effects of changes in the amplitude

FIG. 13.5. *Two lines of chroma signal, showing asymmetry caused by overloading (see text).*

FIG. 13.6. *Two lines of chroma signal, showing the type of distortion produced by overloading in the luminance amplifier. This can also result from misalignment of the chroma bandpass amplifier.*

of the colour-encoded u.h.f. signal applied at the aerial input within its range by progressively decreasing the r.f. and i.f. gain as the signal amplitude rises, and vice versa. This is the normal a.g.c. action, of course.

Fig. 13.6 shows another form of chroma signal distortion which can result from overloading in the luminance amplifier. The distortion is mainly in the yellow and cyan regions of the signal, and if luminance overload is responsible the distortion will fade from the display when the Y signal at the generator is switched off.

Fig. 13.7 shows that the oscilloscope can be used to display the signals going into the chroma detectors. Waveform (*a*) shows the 4·43 MHz reference signal drive which is normally present at both detectors, while waveforms (*b*) and (*c*) show respectively one line of V signal and one line of U signal.

If all is well from the signal point of view up to the chroma detectors, then one can be fairly sure that lack of colour, or a colour fault, has its origin either in the chroma detectors themselves or in the synchronising of the reference signal (by the bursts) or the red detector drive (by the ripple signal).

Lack of reference signal drive to the detectors would delete the colour completely, and if an oscilloscope test shows that the detectors are devoid of drive, then the 'scope Y input should be transferred to the output of the reference generator. If the signal is zero here, too, then the generator is at fault.

REFERENCE GENERATOR TESTS

There are not many things that can go wrong in the generator, the most vulnerable being the crystal and its controlling capacitor-diode. It might be useful, in fact, to disconnect one side of the capacitor-diode before delving too deeply

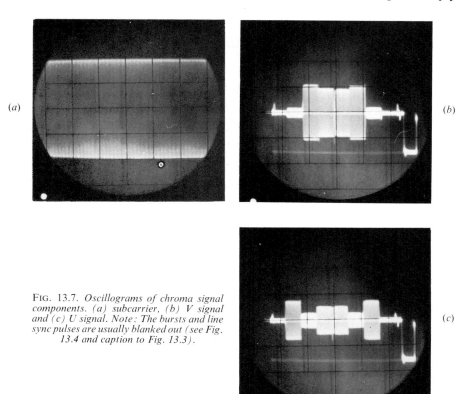

(a)

(b)

(c)

FIG. 13.7. *Oscillograms of chroma signal components. (a) subcarrier, (b) V signal and (c) U signal. Note: The bursts and line sync pulses are usually blanked out (see Fig. 13.4 and caption to Fig. 13.3).*

elsewhere. If this restores the signal, the capacitor-diode should be replaced. The next move would be to check the crystal by substitution, which is not usually a very difficult operation. The transistor could otherwise be in trouble, but one or two voltage tests, using normal servicing techniques, can be used to check this. However, its f_c can sometimes drop for some reason or other, thereby causing oscillation failure at 4·43 MHz. Voltage and current measurements around this transistor will differ from those published when the stage fails to oscillate, even though the transistor itself is in order.

Fig. 13.8 shows the circuit of the Decca reference generator (TR607), its buffer stage (TR608), the phase detector (D613 and D614) and the burst gate (TR606). The buffer stage delivers detector drive signals from its emitter via C644, so if a signal is indicated on the oscilloscope when connected to the output of the generator (e.g. at the collector of TR607), but not at TR608 emitter, then this latter stage or its coupling would be in trouble.

Bursts are applied to the base of TR606 through C629 from the burst tuned circuit (Fig. 13.4), and the transistor is gated by pulses fed in from the burst gate generator (located in the luminance section of the Decca set). Bursts alone are thus applied to the phase detector transformer T1, while a sample reference signal is fed back into the detector via transformer T2. The detector thus produces a control voltage when the phases of the two inputs deviate, and it is this voltage which changes the value of the capacitor-diodes and pulls the generator into step with the bursts, as fully explained in earlier chapters.

It is possible that excessive control voltage from the phase detector, due to a fault condition, could affect the normal operation of the capacitor-diode by applying an abnormally heavy load to the oscillator circuits, thereby cutting off oscillation. This can be proved when displaying the generator output by disconnecting the control voltage input from R647 at the junction of D615 and D616, which are the capacitor-diodes. If this restores the reference signal, then the phase detector circuit, including the two diodes, should be closely investigated.

NO COLOUR SYNCHRONISING

This fault shows as bands of colour drifting in a random manner up or down the screen (provided the chroma channel remains active), and if a colour-bar signal is being used the bars appear correctly in monochrome with their colour drifting away from them. Similarly, Test Card F shows the normal monochrome representation, but with the colours of the picture in the centre circle and of the 'bars' at the top drifting aimlessly.

This means that the reference oscillator is failing to phase-lock to the bursts. There are various possible causes of this fault, including lack of burst input to the burst gate, failure of the gate itself (or its gating pulses), trouble in the phase detector and defective capacitor-diodes (sometimes called 'varicaps'). The first move is to check with the oscilloscope that the gate is, in fact, receiving an input signal, and here we should see the complete chroma signal with bursts either side, rather like the waveform of Fig. 13.3. The bursts are just about visible on this. We are still assuming, of course, that the input to the set is from a colour-encoded generator simulating the colour-bar transmission.

In the Decca set, the bursts are 'plucked' from the collector of the first chroma bandpass amplifier (Fig. 13.4) by the burst tuning, but some dif-

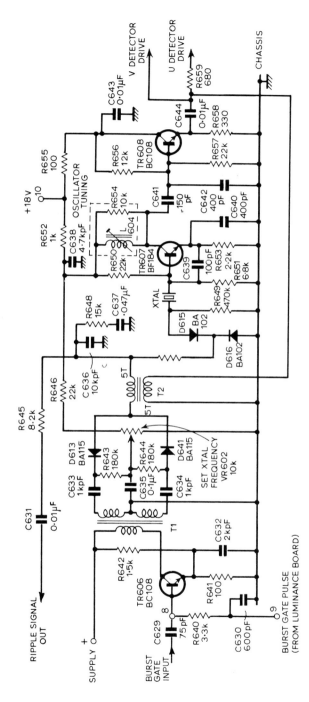

FIG. 13.8. Circuit of the reference generator, buffer amplifier, burst gate and phase detector of the Decca series receivers. Fault-finding procedures and adjustments relating to this section are given in the text.

259

ferences in detail will be observed between sets in this connection. Once we have established that the signal is present at the input of the gate, we can look at the output signal, which should be purely bursts, of course, as shown in Fig. 13.9. Shaped line pulses applied to the gate ensure that it switches on only during the period of the burst, so, if chroma signal is also present at the gate output, attention should be directed towards the biasing of the gating transistor.

In the Decca set, there is a special generator for creating 7 V positive-going bursts at line frequency. It will be seen in Fig. 13.8 that TR 606 (the gate transistor) has no standing forward base bias, which means that it is non-conducting except during the time that it receives the positive-going gating pulses—the switch-on pulses are positive because the transistor is an npn type. Other sets sometimes use shaped and time-adjusted line timebase pulses direct.

In many sets the phase detector control voltage output can be varied up and down by a preset adjustment, which allows the crystal frequency to be established correctly in the first place. If the burst input is removed, the generator

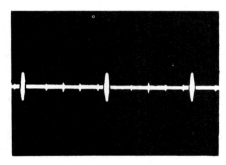

FIG. 13.9. *Bursts after removal of the chroma signal by the 'gate'.*

frequency will come under the complete control of the preset, and if the frequency of the generator changes when the preset is adjusted, we can be sure that the capacitor-diode is working correctly. This is akin to setting the line lock on sets with flywheel-controlled line sync by removing the sync input and adjusting the line oscillator frequency.

If there is no apparent frequency change, then the capacitor-diode and the components directly associated with it should come under scrutiny. If, on the other hand, there is a change in frequency, yet the colour fails to synchronise, the oscillator tuning (L604 in Fig. 13.8) could be out of adjustment. The reference generator should be run as close as possible to the correct frequency with the burst input removed and with the crystal frequency preset at range-centre. When this is done the generator should lock to the bursts immediately they are re-applied.

Another possible fault area which would affect the colour sync embraces the circuit and components concerned with feeding back the sample reference signal from the output of the buffer stage to the phase detector; the main component in this respect in Fig. 13.8 is T2. The two diodes in the phase detector itself, of course, should always be placed high on the list of suspects when a colour sync fault exists.

NO R–Y DETECTOR OUTPUT

Lack of output from the R — Y detector will result in the display of colour bars being predominantly blue, as shown in Plate 5(f). Reading the bars from left to right across the screen under this fault condition they appear as white (normal), yellow (normal), cyan-towards-mauve (instead of cyan), green deeper shade than normal, blue (instead of magenta, dark green veering towards black (instead of red), blue (normal) and black (normal).

This trouble could be caused by failure of the R — Y detector itself or failure of its PAL switching. We can easily check the reference signal drive with the oscilloscope up to the R — Y detector, and the Decca detector and switching circuit sections in Fig. 13.10 reveal the various points involved. The reference signal drive for the R — Y detector emanates from the reference buffer (Fig. 13.8), goes to the right-hand winding of T1 (Fig. 13.10) and thence to either one or other of the two windings on the left-hand side—depending on which of the two diodes, D611 and D612, is conducting. This is where the reference signal phase reversals take place to match the PAL alternate-line switching. The diodes are switched on and off alternately by the action of the bistable, as previous chapters have explained.

If the reference signal is present across the top winding of T2 and across the drive input terminals (at the ends of R637 and R638) of the R — Y detector, then the trouble is likely to lie in the detector diodes themselves or in the signal feeds through and from the PAL delay line. However, to ensure that the trouble does not lie elsewhere in the detector output complex, the oscilloscope can be used to display the R — Y signal (if any) actually leaving the detector. The kind of display expected here when the colour-bar input signal is used is shown in Fig. 13.11. Remember that there are, at least, a couple of possibilities why the colour-difference signal may not be present at the output, even when the detector is working correctly, and they are (a) a short-circuit in the filter capacitor C626 and (b) an open-circuit of the filter inductor L603. These filter components are designed to remove residual reference carrier signal, and if they are functioning properly, the R — Y display should be completely free from trace thickening from this signal. In sets where the inductor is adjustable, the oscilloscope should be connected to the output. The oscilloscope's Y-gain should be increased as much as possible (without reducing the Y bandwidth) and the inductor adjusted to give the smallest amplitude of reference signal on the R — Y signal.

When the R-Y detector switching is out of sync the bars appear as white (normal), off-white (instead of yellow), magenta (instead of cyan), towards-red (instead of green), blue (instead of magenta), dark-green (instead of red), purple (instead of blue) and black (normal).

NO B–Y DETECTOR OUTPUT

Lack of B — Y detector output modifies the display of the colour bars on the picture tube screen in the manner shown in Plate 5(e). Reading the bars from left to right across the screen under this fault condition they appear as white

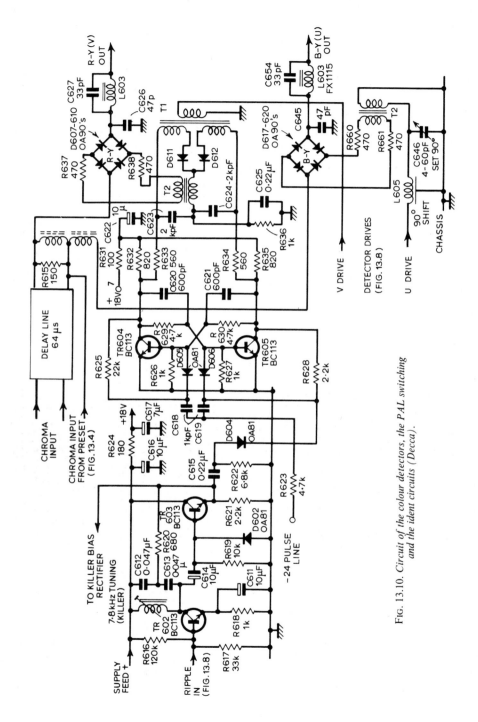

FIG. 13.10. *Circuit of the colour detectors, the PAL switching and the ident circuits (Decca).*

(normal), off-white (instead of yellow), light-green (instead of cyan), a cross between cyan and green (instead of true green), a reddish colour (instead of true magenta), red (normal), black (instead of blue) and black (normal).

This trouble is generally less difficult to locate because the B — Y channel is devoid of the PAL switching complications. However, in some sets the reference signal passes through a 90 degree phase-shift network before arriving at the B — Y detector, as shown in Fig. 13.10. Here it will be seen that the reference signal drive passes through an inductor of a special kind tuned by a 4–60 pF preset capacitor, which adjusts the 90 degree phase displacement. The reference signal is then applied to the right-hand winding of T2, and from the left-hand winding it is fed to the B — Y detector through 470 Ω resistors, matching those at the R — Y detector reference signal input.

Again, the reference signal can be traced with the oscilloscope from the output of the buffer right up to the input of the detector, while the U chroma signal can be traced from the chroma amplifier and PAL delay line up to the chroma input of the detector. The comments made previously in connection with the R — Y reference signal filters apply also the corresponding components at the output of the B — Y detector. Fig. 13.12 shows the B — Y signal derived from the colour-bar input signal.

LACK OF R–Y DETECTOR SYNC

If the R — Y detector is switching out of step with the PAL alternations at the transmitter the colour bars will show on the picture tube screen from left to

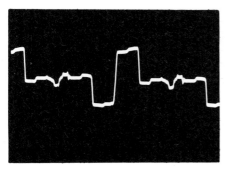

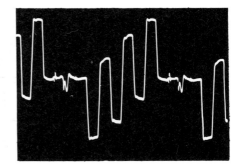

FIG. 13.11. *R — Y signal at the output of the red detector.*

FIG. 13.12. *B — Y signal at the output of the blue detector.*

right as white (normal), yellow (normal), magenta (instead of cyan), orange (instead of green), cyan (instead of magenta), green (instead of red), blue (normal) and black (normal).

This symptom has a fifty-fifty chance of occurring when switching the set on or channel changing should something be amiss with the bistable synchronising from the ripple signal, or should the ripple signal itself be missing or low in amplitude. A practical way of checking the performance of this synchronisation is to interrupt the input signal from the generator (or the aerial if the radiated test bars are being used) to the aerial input socket of the set. If all is well, the correct colour will occur each time the signal is re-applied after being removed. If the correct colours occur only five times out of ten, with the incorrect colours (as just detailed) occurring the other five times, then the bistable is almost certainly without ripple ident signal. If *incorrect* colours occur less frequently, say, two times out of ten, the R — Y switching sync is weak.

Some sets feature a sort of R — Y detector sync threshold preset control which needs to be carefully adjusted to secure optimum sync performance. The procedure for adjusting this is given in Chapter 14.

In the Decca circuit (Fig. 13.10) ripple signal from the phase detector (Fig. 13.8) is applied to the base of TR602 and developed in a high-Q circuit, tuned by L602 in the collector. This signal (at 7·8 kHz) is delivered at low impedance by TR603 and applied as ident signal to the bistable consisting of TR604 and TR605 in conjunction with the diodes D605 and D606, which endow the circuit with 'binary counter' attributes.

The bistable is normally switched by -24 V line pulses applied through R624, C618 and C619. The oscilloscope can be used to investigate the ripple signal at both TR602 collector and TR603 (emitter-follower) emitter, but to avoid confusing indications from the line pulses and the bistable switching action resulting from these pulses, it is best to isolate the ident signal from the bistable by disconnecting, say, diode D604 from the junction of C615 and R622 in the circuit of Fig. 13.10.

It will be seen that the ripple signal at TR603 emitter is used, too, as colour-killer bias after rectification, so trouble in the ripple circuit could also give rise to lack of chroma by failing to bias-on the chroma bandpass amplifier (Fig. 13.11).

The actual switching action of the bistable itself can be observed on an oscilloscope, after adjusting the sweep control to a suitable frequency. It is then possible to tell whether or not the trouble lies in the bistable, in its switching input (from line pulses) or in its synchronising (from ident signal).

When it is possible to switch from PAL to NTSC, and vice versa, on the colour-bar generator (as with the Körting 82512), the ident circuits can be tested simply by switching alternately over the two standards. If the circuits are working correctly the PAL switch in the set will phase-lock on to the correct line each time the generator is switched to PAL. Of course, when the generator is switched to NTSC a PAL set will switch to monochrome operation because the non-swinging burst of the NTSC system will fail to produce a ripple signal for biasing-on the chroma bandpass amplifier channel.

It is desirable to check the switching action at all the available combinations of chroma signal, such as R — Y alone, B — Y alone and B — Y with R — Y.

PAL DECODER ALIGNMENT

The PAL decoder is here assumed to be that network, including the PAL delay line and adder/subtractor matrix, between the chroma bandpass amplifier and the synchronous colour detectors, the alignment of which can be performed one way in conjunction with a colour-bar generator incorporating facilities for switching on and off the chroma signals separately.

When correctly adjusted, the decoder should not show an output from the U arm (e.g. that feeding the B — Y detector) when the U chroma signal is switched off at the generator. An oscilloscope with a high Y-gain is required to indicate the signal at this point, for its amplitude is not likely to be very high. Wrongly adjusted, a display like that in Fig. 13.13 will be obtained with the oscilloscope's Y-input connected to the chroma input of the B — Y detector.

This shows the presence of V chroma signal, and the decoder matrix should be adjusted until the chroma signal fades away, as shown, for instance, by

Fig. 13.13. *V chroma signal at the output of a maladjusted PAL decoder (see text).*

Fig. 13.9. It is desirable to run through the test again but this time with the V chroma signal switched off at the generator and the oscilloscope connected to monitor the U chroma signal at the R — Y output of the decoder.

To secure the zero-chroma condition of Fig. 13.9, it is necessary carefully to adjust any reference frequency matrix or phase correcting presets that the set under adjustment may feature. For example, VR601 in Fig. 13.4 is an adjustment of this kind.

Incorrect PAL matrix adjustments often gives the symptom of Hanover bars.

SYNCHRONOUS DETECTOR ADJUSTMENT

Each colour detector can be adjusted for maximum amplitude of colour-difference signal as indicated by an oscilloscope or valve voltmeter connected to the output of the corresponding colour-difference amplifier when the generator is modulated only with that particular colour-difference signal. For instance, by switching on only the U chroma signal at the generator the B — Y detector can be adjusted to yield the maximum B — Y signal. A similar adjustment can be carried out with the oscilloscope or valve voltmeter connected to the R — Y amplifier output when the V chroma signal alone is modulating the generator.

It is also possible to observe the adjustments subjectively on the colour-bar display at the picture tube screen in terms of saturation, first on the V chroma signal and then on the U chroma signal. Displays as shown at (*e*) and (*f*) respectively in Plate 5, will be obtained when the adjustments are correct.

ALIGNMENT OF THE COLOUR DETECTOR PHASING

With a generator whose standard can be switched over PAL and NTSC, the alignment of the colour detector phasing can be checked as follows.

B — Y Detector

1. Switch the generator to the NTSC standard colour bars.
2. Disable the colour killer; that is, bias the chroma controlled amplifier so that it continues to function independently of the missing swinging burst of the NTSC standard.
3. Switch off the U chroma signal at the generator.

If the phasing is correct at 90 degrees, the red bar displayed on the screen will appear colourless (grey). If there is a bluish or greenish tint the phasing is in error, and the phasing inductor in the reference signal drive to the B — Y detector should be adjusted to obtain the correct result.

R — Y Detector

1. Switch off the V chroma signal at the generator.
2. Re-arrange the decoder outputs so that the U chroma signal appears at the inputs of both colour detectors.
3. Connect the Y input of an oscilloscope to the output of the R — Y detector and check that zero signal voltage is present. If signal is indicated, the phasing of the R — Y detector should be adjusted to give a voltage null.

It should be noted that if the set under test has a 90 degrees phase shift inductor inserted in the B — Y detector (as in Fig. 13.10), the R — Y detector check should be performed first.

ALIGNMENT CHECK BETWEEN SATURATION AND CONTRAST

The colour-bar encoded generator can also be used to check that the correct conditions exist between the Y and chroma channels in the following way:

1. Switch off the green gun.
2. Unlock the vertical hold so that the field blanking period is visible about half way down the screen.
3. Switch the generator to provide the complete colour-bar signal.

Now, if the correct conditions exist, the green bar will have the same black level as the field blanking period.

ADJUSTING THE COLOUR-DIFFERENCE CIRCUITS

By feeding the colour-bar signal into the set it is possible to adjust and balance, if necessary, the signal amplitudes at the picture tube, irrespective of whether colour-difference or primary-colour drive is adopted, as explained in the sections that follow.

Signal Tracing

The colour-difference signals resulting from a colour-bar generator input can be traced through the appropriate colour-difference preamplifiers and output stages by means of an oscilloscope. The R — Y and B — Y signals (Figs. 13.11 and 13.12 respectively) should, of course, increase in amplitude when tracing from stage to stage from the colour detectors to the picture tube grids.

The G — Y signal is derived from a process of matrixing the R — Y and B — Y signals, as we are now very well aware, and this signal, as obtained from the standard colour bars, is shown in Fig. 13.14.

Checking and Adjusting the G — Y Matrix

If the generator is modulated only with the U chroma signal, for instance, and the oscilloscope is connected to indicate the proportion of B — Y signal fed to the G — Y matrix, we should get a display rather like that in Fig. 13.15, where the top trace is the B — Y signal proper and the bottom trace shows the proportion of that signal delivered by the G — Y matrix. The phase, it will be seen, is inverted.

In a similar manner, we can switch off the U and switch on the V chroma signal at the generator and obtain a display like that in Fig. 13.16, where the top trace is the R — Y signal proper and the reduced amplitude, bottom trace is the inverted R — Y signal, also delivered by the matrix. When both B — Y and R — Y signals are present, the bottom signals of Figs. 13.15 and 13.16 combine to form the real G — Y matrix output signal shown in Fig. 13.14. In practice, of course, the three colour-difference signals are fed into colour-difference output stages, from which they emerge as colour-difference drives for the picture tube grids or, alternatively, the colour-difference signals are fed into a further circuit (also called a matrix) along with the Y signal to yield primary-colour picture tube drive signals.

Figure 13.17 shows the colour-difference stages of the Decca series receivers which, in common with many other sets, use transistors for the preamplifiers and valves for the output stages (the pentodes of the valves being the amplifiers and the associated triodes the clamps). The R — Y signal is fed to the base of transistor TR609, while the B — Y signal is fed to the base of transistor TR611. Both of these transistors deliver the amplified signals at their collectors.

The matrixing action occurs by TR609 and TR611 feeding proportions of the R — Y and B — Y signals from their emitters to the emitter of the common-base

G — Y preamplifier transistor TR610, via the matrixing network comprising R673, R676 and the matrix adjusting preset VR604. When the matrix is properly adjusted, the correct proportions (the signs of the signals are automatically established correctly by the nature of the circuit) of the R — Y and B — Y signals are applied to TR610 emitter and the correct G — Y signal appears at its collector.

By displaying on an oscilloscope the R — Y and B — Y signals in isolation at the collector of TR610 (for instance by switching off at the colour-bar

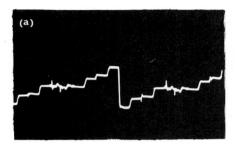
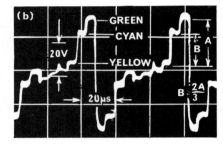

FIG. 13.14. *(a) G — Y signal at the output of the G — Y matrix. (b) Signal component amplitude ratios for correct adjustment.*

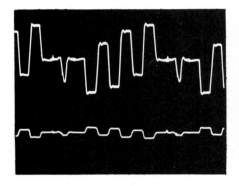
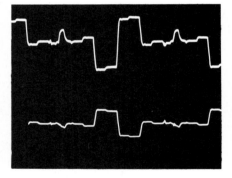

FIG. 13.15. *Bottom trace is the proportion of the B — Y signal from the top trace, inverted, required along with a suitable proportion of R — Y signal to form the G — Y signal (see Fig. 13.16).*

FIG. 13.16. *Bottom trace is the proportion of the R — Y signal from the top trace, inverted, required along with a suitable proportion of B — Y signal (Fig. 13.15) to form the G — Y signal.*

generator that signal not being displayed), it is possible to adjust the matrix preset to secure the required proportion of each one to yield the correct G — Y signal. Theoretically, the required $-(G — Y)$ signal is equal approximately to $0{\cdot}51(R — Y) + 0{\cdot}19(B — Y)$, so that the matrix should, on this basis, be adjusted to yield red and blue colour-difference signals in the ratio of $0{\cdot}51{:}0{\cdot}19$. In practice, however, the initial weightings of the U and V chroma signals can modify this so, in the advent of the maker's information on setting-up the matrix not being available, the best plan is to display the G — Y signal at the output of the G — Y preamplifier stage (e.g. the collector of TR610 in Fig. 13.17)

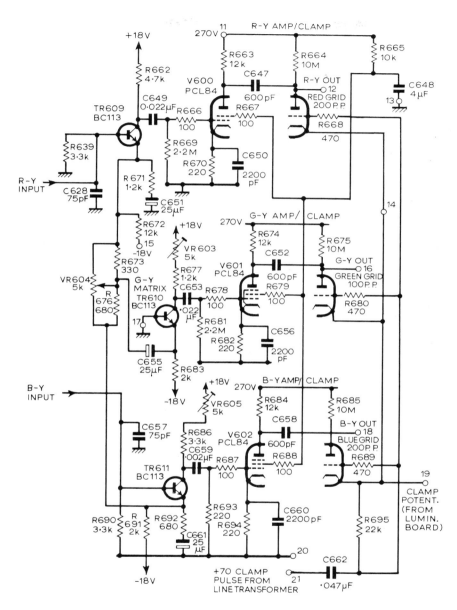

Fig. 13.17. $G - Y$ matrix and the colour-difference channels (Decca). Fault-finding procedures and adjustments relating to these channels are given in the text.

on an oscilloscope when the colour-bar signal is applied to the input of the set, and then adjust the G — Y matrix preset to give the correct waveform, as shown in Fig. 13.14(a). Figure 13.14(b) highlights the relative amplitudes of the components of the G — Y signal when the matrix is correctly adjusted.

Checking and Adjusting the Colour-Difference Drives

It is not uncommon to find presets in the collectors of the G — Y and B — Y preamplifier transistors. These make it possible to adjust the green and blue colour-difference drives to the control grids of the corresponding output valves relative to the fixed drive to the R — Y output valve. Such presets are VR603 and VR605 in the Decca circuit, Fig. 13.17.

The drives can be checked by displaying the signals produced by the colour bars at the control grids of the output pentodes on an oscilloscope and, if necessary, those to the green and blue output stages can then be adjusted to give the required colour-difference drives from the pentode anodes to the picture tube grids. These adjustments are generally given in the maker's service manual, where slight differences in detail are likely to be observed between different makes and models.

However, if each output stage is designed for the same gain, which is often the case with colour-difference picture tube drives, and if the maker's data is not at hand, then there is a method of making the adjustment, by using an oscilloscope to display the drives at the grids of the pentodes separately when the set is receiving a colour-bar signal. This method is to regulate the G — Y drive preset until the G — Y peak-to-peak voltage is about 0·6 times the fixed R — Y peak-to-peak voltage and the B — Y drive preset until the B — Y peak-to-peak voltage is about 1·2 times the fixed R — Y peak-to-peak voltage. Also see Fig. 13.14(b).

CLAMPING ACTION

The foregoing checks and adjustments, therefore, will soon reveal faults or shortcomings in the colour-difference stages, the correction of which follow normal monochrome servicing practice.

The oscilloscope can also be used to check the action of the clamps at the output of the colour-difference amplifiers. The clamping pulses can be observed at the cathodes of the triodes, for instance, in Fig. 13.17, while the effect of the clamping—when it is working properly—will be seen on the colour-difference signals delivered at the anodes of the colour-difference output pentodes.

Y CHANNEL CHECKS

As already mentioned, the signal in the Y channel—from the video detector to the picture tube cathode or primary-colour signal drive matrix—is virtually the same as that in a monochrome set from the vision detector onwards, so the procedures adopted in black-and-white set servicing can also be applied to the Y channel in colour sets.

There are extra items and circuit differences in the Y channel compared with the straightforward monochrome video channel, though, and these include preamplifier stages, the luminance delay line and tuned filters, notably the 4·43 MHz subcarrier notch filter. An extra complication exists in some models in terms of a transistor switch which automatically disconnects or damps the subcarrier filter when the transmission is in black-and-white, thereby ensuring that the monochrome display retains optimum definition (see Fig. 13.19). All these factors have been adequately covered in past chapters.

However, when a colour-bar generator is available, the stepped Y waveform produced by this with the chroma switched off (Fig. 13.2b), is ideal for checking the grey-scale performance. The actual results can be appraised subjectively by observing the grey-scale pattern on the screen or objectively by the use of an oscilloscope. When an oscilloscope is used to monitor the signal in the Y channel, overloading shows as in Fig. 13.18. Here the waveform is badly crushed at the dark-grey and black end. This results in severe compression of the sync pulses, which could affect the sync performance as well as the grey-scale rendering.

This sort of trouble would point to overloading in the r.f./i.f. amplifier channel (by insufficient a.g.c. voltage, for instance), when the distortion is on the

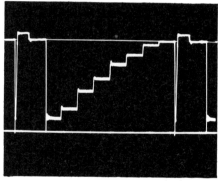

Fig. 13.18. *Crushing of the luminance signal due to overloading in the Y channel. Compare with the oscillogram in Fig. 13.2(b).*

waveform monitored at the video detector output. If there is little or no distortion at that point, yet distortion on the signal along the Y channel, or at the cathodes of the picture tube, the overloading would then, of course, be taking place in the Y amplifiers, and by monitoring the signal from stage to stage along the channel the fault area would soon be revealed. Typical troubles are faulty valves and transistors, and incorrect biasing of the preamplifier transistors or the Y output valve.

On some generators the Y signal (at video) is also available, and, if necessary, this can be applied to the input of the Y channel, or in turn to the various stages, as a means of isolating a faulty section.

When the picture tube screen itself is used to assess the grey-scale performance, it is a good idea to slip the vertical hold so that the horizontal band corresponding to the field blanking is visible towards the centre of the screen. The degree of black level between the field sync pulses and the field blanking will give a good indication of pulse clipping. However, an oscilloscope is necessary to determine which stage is responsible for the bad performance.

TUNING THE 4·43 MHZ NOTCH FILTER

One way of adjusting this filter is by applying a signal at the aerial input from a colour generator with the Y signal switched off, and then observing the sub-carrier on an oscilloscope connected to the input of the luminance channel. The Y gain of the oscilloscope should be adjusted to give the greatest vertical deflection possible, and then the Y input of the oscilloscope should be transferred to the collector of the transistor in whose emitter circuit the notch filter is connected. Normally, the notch filter will afford a very high degree of attenuation to the subcarrier signal, so a considerable oscilloscope Y gain will be required at 4·43 MHz to observe it. The filter should be adjusted with a non-metallic trimming tool for minimum subcarrier output.

Alternatively, the subcarrier signal direct from the generator (if available separately) can be applied to the input of the luminance amplifier, and the adjustment made with the oscilloscope as just described. This method ensures that a strong subcarrier signal is fed into the luminance channel, but it does require a generator with a subcarrier output facility. A typical Y channel first preamplifier stage is shown in Fig. 13.19 with the generator and oscilloscope connected for filter adjustment.

OTHER FAULT ASPECTS OF THE LUMINANCE CHANNEL

It is a simple matter to trace the signal from the input to the output of the Y channel with an oscilloscope—a procedure that will soon bring to light any faulty section responsible for complete luminance failure.

Faults sometimes occur which have no direct parallel to those in the video channel of monochrome sets. One is severe dot-pattern on the screen, and this invariably indicates misalignment of the 4·43 MHz notch filter. When out of adjustment, it is possible for this filter to put a deep trough in the Y channel response at a frequency removed from 4·43 MHz, the effect of which can be impaired picture definition, depending on which direction the notch is out of tune and on whether the trough resulting from it falls in the luminance passband.

Another fault is horizontal misregistration of the luminance with respect to the colour on the picture itself. This sort of symptom invariably implies that the Y information is failing to arrive at the picture tube at exactly the same time as the colour information. The luminance delay line slows down the Y signal to give the correct Y and colour timing, so if this line is faulty or if the connections to it are badly made, this symptom is likely to crop up.

It is also noteworthy that misalignment of the i.f. stages or the chroma bandpass amplifier can upset the timing, even though the luminance delay line and its connections are correct. This is because alteration in alignment of the tuned stages will alter the time taken by the signals to pass through them. The characteristics of the luminance delay line are, of course, patterned in relation to the signals operating in correctly aligned circuits.

A bad line or terminations to it can incite reflections, giving symptoms of 'ringing' on the picture. This trouble is described in Chapter 8.

Luminance overloading can also develop because of incorrect preamplifier transistor biasing, the effect of which can be aggravated when the brightness control, of the type varying the standing current in the luminance channel, is operated.

Many luminance amplifiers use a.c. coupling at their inputs by reason of a coupling capacitor connected from the video input to the base of the first

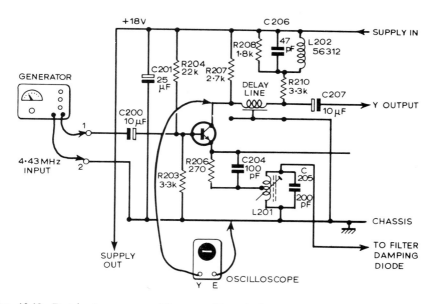

FIG. 13.19. *First luminance preamplifier stage (Decca), showing the set-up for adjusting the 4·43 MHz notch filter. Note: During monochrome reception trap L201 is fully damped by a diode (not shown).*

transistor (e.g. C200 in Fig. 13.19). This calls for some form of clamping or d.c. restoration (described in Chapter 8). If this fails, the sync and black-level performance will be substantially impaired.

Details, procedures and possible fault conditions related to the Y output drive circuits and their adjustments to the cathodes of the picture tube are fully dealt with in Chapters 8, 12, 14 and 16, while Chapter 9 covers a.g.c. and the sound channel, the faults and servicing procedures of which have very much in common with those of monochrome sets.

OTHER SUBJECTIVE COLOUR BAR CHECKS

Apart from indicating how well the decoder and synchronous detectors are adjusted, the colour bars displayed on the screen of the picture tube can inform the technician of possible tube gun and associated control circuit faults. For instance, with the blue gun defunct, the bars would appear from left to right as yellow (instead of white), yellow (normal), green (instead of cyan), green (normal), red (instead of magenta), red (normal), black (instead of blue) and black (normal). With the red gun defunct the bars would appear as cyan (instead of white), green (instead of yellow), cyan (normal), green (normal), blue (instead of magenta), black (instead of red), blue (normal) and black (normal). With the green gun defunct the bars would appear as magenta (instead of white), red (instead of yellow), blue (instead of cyan), black (instead of green), magenta (normal), red (normal), blue (normal) and black (normal). Also see Plate 5(*a*), (*b*) and (*c*).

With only a single gun working (the other two defunct, switched off or for some other reason failing to produce illumination), only the bars carrying the primary colour of the active gun would appear. For example, with the red gun only running, red would appear at the positions of white, yellow, magenta and red; with the green gun only running, green would appear at the positions of white, yellow, cyan and green; with the blue gun only running, blue would appear at the positions of white, cyan, magenta and blue.

The guns working to produce *white* are red, green and blue; to produce *yellow*, red and green; to produce *cyan*, blue and green; to produce *green*, green only; to produce *magenta*, red and blue; to produce *red*, red only; to produce *blue*, blue only; while black occurs when all the guns are quiescent. It will be recalled that yellow, cyan and magenta are the complementary colours, which is why they each need a mix of two colours; they are complements of the missing colour required to make white. Thus, yellow is the complement of blue, cyan of red and magenta of green (see Chapter 16).

The chroma signals are $-(B - Y)$ to give yellow; $-(R - Y)$ to give cyan; between $-(B - Y)$ and $-(R - Y)$ to give green; between $+(B - Y)$ and $+(R - Y)$ to give magenta; $+(R - Y)$ to give red; and $+(B - Y)$ to give blue.

Use of Test Card F

For subjective colour appraisal the first few top lines of Test Card F, carrying the colour bars (see Chapter 11), can be employed either by slipping the vertical lock or reducing the height. It is also possible to use these first few lines for oscilloscope displays by triggering the oscilloscope's timebase by the field scan.

FREQUENCY RESPONSE APPRAISAL BY THE CROSSHATCH DISPLAY

The crosshatch pattern delivered by colour television pattern generators is ideal for assessing the overall vision frequency response in terms of how sharply the vertical lines of the pattern are defined on the screen. Poor alignment is revealed by overshoot (ringing) and slurring of the edges of the vertical bars.

The vertical bars can also be used to check the luminance amplifier for ringing effects, while hum symptoms are highlighted by this sort of display.

CHAPTER 14

Servicing in the Field

MOST SERVICING takes place in the customer's home. Owners are always reluctant to be parted from their sets—especially colour ones in which they have invested several hundred pounds—even for a few days, while viewers of rented sets are probably protected against loss of service through set breakdown by their rental contracts; meaning that if such a set is taken away for repair an equivalent replacement has usually to be left behind.

Moreover, although not always particularly large, a colour set is a heavy integration of sophisticated electronics and sometimes polished furniture, embodying a costly and critically adjusted colour picture tube, thus being vulnerable inside to maladjustment and outside to scratching, and generally needing two people to carry it.

Also, since a set is best degaussed and finally adjusted in the position from which it is to be viewed, two additional fundamental operations are made necessary by taking the set from the home for repair—resetting to the conditions of the workshop during the repair and resetting again in the home after the repair. Quite a lot of time, frustration and expense can thus be saved by handling the repair in the home.

Fortunately, designers of colour sets have been made aware of the need for field servicing and have designed accordingly, on the plug-in unit or module basis, with an eye sharply focussed on ease of replacement and accessibility. This is necessary when one considers how much more complicated a colour set is compared with a monochrome counterpart.

The technician should, therefore, carry with him a set of units and/or modules for the sets he is responsible for servicing, to diagnose in the field which unit or module is at fault, to extract this and fit a replacement and either to repair the faulty one at leisure back in the workshop or to return it to the maker for a reconditioned replacement. This technique has much in common with car servicing; and it certainly speeds up colour set servicing, making it possible to clear quite complex troubles in the field which, in the ordinary way, would have demanded workshop attention.

Quite often the defective unit or module is returned to the maker for repair and/or replacement, but to avoid television servicing becoming too much of an automated process many technicians prefer to trace and clear the section faults in the workshop, and for this it is not uncommon to find a 'test jig' installation into which the defective section can be plugged and operated for dynamic tests and measurements.

SERVICING IN THE FIELD

ASPECTS OF FIELD SERVICING

Of course, the success of field servicing of this kind depends on two prime factors: (*i*) how well the technician is equipped with replacement sections and (*ii*) how speedily he is able to diagnose the fault and pin-point the section responsible for the trouble. There are times, of course, when workshop attention is essential, such as when a picture tube needs replacing (it is not really a good idea to undertake this rather involved job in the home—although some technicians get away with it quite well), and when the overall alignment of the set requires detailed attention (this, too, could be handled in the home if the viewer has no qualms about his lounge being turned temporarily into a test laboratory!)

Field servicing thus consists of the following.

1. Establishing the actual fault symptom.
2. Being sure whether the symptom is the result of misadjustment of the user's and preset controls or of a real fault.
3. Locating the fault to a specific area of the set and finally to a specific module or unit (e.g. fault diagnosis).
4. Replacing the faulty section.
5. Re-adjusting the set for optimum performance to the viewer's complete satisfaction.

Chapter 12 provides information on points 1, 2 and 3, while Chapters 6 and 8 and the maker's service manual give all that needs to be known about points 4 and 5.

Speedy field servicing is closely geared to how quickly one is able to identify the various sections and units inside the cabinet and to relate them to specific fault symptoms. The fault diagnosis chart in Chapter 12 should help in the latter respect. When the back is taken from a colour set a fair number of sections will be recognised in terms of their equivalent in monochrome sets. Immediately apparent, for example, will be the line output stage, booster diode and e.h.t. section; also the i.f. strips, the sound channel and possibly the field timebase section. Some of the other sections will be seen to be common only to colour sets, including the conglomeration of components on the tube neck, the convergence and control panel and the sections of the decoder.

Technicians concerned with the servicing of only one type of colour set—such as a rental set with a common chassis but coming in a variety of cabinets—will very quickly become accustomed to the layout of the sections inside—also, of course, to their purpose and, indeed, to many of the so-called 'stock faults' which gradually come to light after a model has been active in the field for several months.

Technicians who are expected to handle the servicing of all makes will have a greater number of sets to study, and although variety will probably make their job more interesting, it will still be somewhat less involved than that of technicians of a decade or two past, who were called upon to handle multitudes of sets of different makes, models and designs. This is no longer the case, for progressive amalgamations of leading manufacturers has resulted in the production of only a handful of new sets each year which essentially differ only in cabinet, banner and model number.

Figure 14.1 shows the inside of a RBM 25 in. dual-standard table model. The elliptical speaker can be seen in the top left-hand corner of the cabinet, with the convergence and control panel slightly to the right at the rear edge of the cabinet. The unit vertically positioned on the left-hand side of the cabinet is the i.f. strip,

FIG. 14.1. *Inside view of a Rank-Bush-Murphy dual-standard set, showing layout of the various units (courtesy, The Rank Organisation).*

to the bottom left of which is the power unit, then to the right of this are located the audio and a.g.c. units, while the unit in main view at the bottom of the cabinet is the decoder carrying the PAL delay line. Next to the decoder is the luminance and colour-difference amplifier section, while the large, L-shaped unit on the right of the cabinet embodies the field and line timebase and e.h.t. circuits.

The rear of the tube looks very similar to a monochrome tube; but it differs in terms of neck components (see Chapter 16, for example) and by the metal band it carries round the edge of its screen.

The set shown is hybrid, comprising both valve and transistor circuits, and some of the main sections just mentioned are illustrated in detail in Figs. 14.2, 14.3, 14.4, 14.5, 14.6 and 14.7.

The 405/625 standard-change switch can be seen running along the length of the i.f. unit in Fig. 14.2. The two controls in the top right-hand corner are for presetting the contrast on 405 and 625 lines, while the adjustment on the lower edge is for regulating the system for local or distant stations. This unit is fully transistorised.

The audio unit in Fig. 14.3 has its two output transistors adjacently mounted in heat-sinks on the top metal panel and its amplifier and driver transistors on

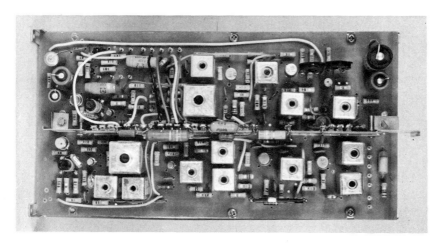

FIG. 14.2. *Detailed picture of the Bush CTV25 i.f. unit, showing the standard-change switch running along the middle of the unit (courtesy, The Rank Organisation).*

the printed-circuit board below. The two presets in the middle of this board are for setting the bias and balancing the output transistors for minimum crossover distortion and optimum audio quality.

The decoder unit in Fig. 14.4 is also fully transistorised, and the large, rectangular, box-like component is the PAL delay line. The subcarrier signal is applied via the coaxial cable at the centre of one of the long sides of the board

FIG. 14.3. *Audio unit (Bush CTV25). Notice the output transistor mounted in heat sinks on the vertical metal plate (courtesy, The Rank Organisation).*

(at the top of the picture), while the composite video signal is fed in at two pins on the opposite side of the board.

The luminance and colour-difference amplifier unit in Fig. 14.5 carries both valves and transistors. The transistors are mostly concerned with reference signal generation and V detector drive switching, while to the valves are delegated the jobs of luminance and colour-difference signal amplification. The picture shows the reference generator crystal in the top left-hand corner of the board with the two diode-capacitors, locking its phase from the bursts, close by.

The power unit in Fig. 14.6 has much in common with any other kind of power unit. The thermistor which is visible is for heater stabilisation, but there is

FIG. 14.4. *Decoder unit, showing the PAL delay line (Bush CTV25) (courtesy, The Rank Organisation).*

FIG. 14.5. *Hybrid luminance and colour-difference unit (Bush CTV25) (courtesy, The Rank Organisation).*

another one in the unit having a positive temperature coefficient for the auto-degaussing system. The fuse holders reading from top to bottom carry a 1 A fuse for the transistor supplies, a 2·5 A fuse for the h.t. supplies and a 5 A fuse for the mains input.

The timebase and e.h.t. unit in Fig. 14.7 has a strong characteristic resemblance to the same sections in monochrome sets. The two horizontally mounted valves at the top are the 25 kV e.h.t. regulator (nearest to the screen) and the 25 kV e.h.t. rectifier. The line output transformer is directly below, and the highly insulated winding for the e.h.t. rectifier heater is clearly visible making two turns round a limb of the transformer core.

BEWARE OF X-RAYS

Nearby preset controls are for width, e.h.t. regulator current and focus (both standards). The top section of this unit is normally tightly screened, as shown in Fig. 14.1. Although x-ray radiation is well within the safe dose-rate when the set is working correctly and properly screened it can, nevertheless, rise above the safe intensity close to the e.h.t. rectifier and regulator valves inside the screen,

Fig. 14.6. *Power unit (Bush CTV25). The thermistor visible is for heater chain stabilisation. A further positive-temperature coefficient thermistor is used for the auto-degaussing circuit (courtesy, The Rank Organisation).*

Fig. 14.7. *Timebase and e.h.t. unit (Bush CTV25) (courtesy, The Rank Organisation).*

especially when a fault condition exists in the e.h.t. supply circuits or associated valve, and it is thus imperative to ensure that the screens are all securely fitted after a servicing operation. The radiation can also rise ten-fold above normal if the e.h.t. circuits are incorrectly adjusted or if a fault exists which causes the e.h.t. to rise substantially above its normal 25 kV.

The line output valve, booster diode and focus e.h.t. rectifier are all positioned on the far side of the screen, visible in Fig. 14.1. The focus rectifier is the horizontally mounted valve. There is also a fourth, two-section, valve in this compartment just visible in Fig. 14.1, which is the line oscillator and flywheel sync reactance valve. The presets just in front of it are for line-locking on 405 and 625 lines.

The two valves on the right of Fig. 14.7 and on the left of the timebase unit in Fig. 14.1 are concerned with the field timebase, the large one being the vertical output valve and the short one the inverter and sync amplifier. The rear of the vertical hold control is just visible in the far right-hand corner in Fig. 14.7.

BRC SETS

Another noteworthy example of unit board plug-in design is seen in the British Radio Corporation's range of dual-standard all-transistor sets based on the 2000 series chassis. Here ten plug-in unit boards are arranged to slide easily in and out of a metal chassis frame, as shown in Fig. 14.8. This allows the removal and replacement of a faulty board in a matter of seconds. The chassis also slides out of the cabinet on runners for speedy accessibility of the boards. This chassis incorporates the Thorn 'Jelly Pot' e.h.t. rectifier assembly, which can be seen mounted on top of the picture tube in the photograph. Figure 14.9 shows a similar, earlier, chassis with the ten main circuit boards removed from the

Fig. 14.8. *View of the chassis of the BRC 2000 series. This model uses plug-in module boards—see Fig. 14.9. The single-channel 3000 series is designed similarly (courtesy, Thorn Electrical Industries).*

metal frame and these are shown placed relative to their mounting positions. These boards (or modules) have been referred to extensively throughout this book, and almost all of them have been illustrated and described separately.

CONTROL IDENTIFICATION

After all the various sections have been adequately identified, the next move is to identify the internal preset controls. Quite a few of these, of course, will have their equivalents in monochrome sets, such as the vertical and horizontal holds,

FIG. 14.9. *The ten main module boards of the BRC 2000 series chassis removed and placed in their relative positions round the frame (courtesy, Thorn Electrical Industries).*

the width and height presets, linearity and so forth, but there are a number of others which appear only in colour sets. The position of these and the effect they have on the performance of the circuits must be fully understood.

All sets have a control panel for dynamic convergence. This is usually hinged so that it can be swung-up above the top of the cabinet, as in Fig. 14.1, or else it is detachable on an extended cable so that the controls can be adjusted while the screen is observed. The correct way to adjust the dynamic convergence is dealt with in Chapter 6, and more information is given later in this chapter.

The convergence control panel might also carry other controls. The panel of the RBM sets, for instance, carries the field timebase height and linearity controls, the picture tube first-anode presets for grey-scale tracking, and the red, green and blue gun switches. The luminance drive presets, also for grey-scale tracking, are located on a small panel fixed to the base holder of the picture tube, together with a focus preset.

It will be useful at this stage to consider a typical procedure (RBM) for initial installation, for not only will the field technician be expected to perform this on each new installation, but he will also almost certainly have to run through it when called out on a repair to establish points 1 and 2 referred to previously.

TYPICAL INSTALLATION PROCEDURE

Assuming that the receiver has been carefully checked in the dealer's workshop in accordance with the setting up instruction, the following procedure should be adopted on initial installation.

1. Check mains adjustment (note that on some sets—Rank sets for instance—this may require removal of the power unit).
2. Select a suitable position for the receiver in the customer's room so that the screen is shielded from direct light from lamps or windows. Avoid positioning the receiver close to any large metal object such as a hot water radiator etc., otherwise stray magnetic fields may disturb the receiver convergence and purity.
3. Remove the cabinet back, and connect the receiver to the mains supply. *Observe the usual live chassis precautions.* Switch on the receiver and allow a warm up period of approximately 10 minutes before making any adjustments.
4. Check the purity by looking at the red, green and blue fields in turn and carry out the degaussing procedure (if required) as follows. Connect the degaussing coil to the mains supply, and degauss the complete receiver including the shadowmask and chassis. Remove the degaussing coil from the vicinity of the receiver before switching off the coil.
5. Connect both aerials and check that all the local stations are received satisfactorily and that the tuning on all buttons is correct. Adjust the brightness control, the 405 and 625 presets and the contrast control for a correct picture. The local/distant control may need adjustment if a noisy picture or cross-modulation is observed.
    Note: An aerial system which is satisfactory for monochrome reception may not be entirely suitable for the reception of colour transmissions due to multipath reception, ghosting, insufficient bandwidth etc., and, while this may have little or no effect on monochrome reception, it could produce any of the following symptoms on a colour transmission: colour distortion, grain on colour, weak colour irrespective of the setting of the colour control. Under these conditions adjustment to or resiting of the aerial system may be needed.
6. Switch the receiver to a 625 colour transmission, set the colour control to the centre of its travel and adjust the preset auto-chroma gain control (Rank sets, for instance) for a correctly saturated colour picture.
7. Remove the aerial input and connect in its place a pattern generator to provide a crosshatch grid pattern of appropriate line standard. Set the

brightness control so that the entire raster is visible. Adjust the purity ring magnets for a completely red field and then check the purity of the green and blue fields. Adjust the radial and lateral static convergence magnets if required. Repeat the purity adjustments if a large static convergence error is corrected.

8. *Check dynamic convergence*—switch the pattern generator and receiver to 405 lines and check the purity, static and dynamic convergence.

9. Replace the back cover, re-connect the aerials and instruct the customer in the operation of the receiver, particularly the slight tuning adjustment that may be required for the best results on colour transmissions.

SETTING-UP PROCEDURE

If the above procedure fails to provide correct monochrome and colour reception, then the set will either be faulty or its main adjustments will be in error. It might well be, for example, that while the purity adjusts properly, the convergence or grey-scale tracking is wildly out of trim. In this event, it would be desirable to run right through the full setting-up procedure as detailed by the maker in his service manual, and at some point in this procedure it should become evident whether the incorrect working of the set is, in fact, due to mal-adjustment or a definite fault.

It is not proposed to run through the setting-up procedure of one particular set, for the various models are best adjusted in their own special ways. However, the following points will give a guide as to the order in which the adjustments can be performed and of the observations to make while they are being made.

Grey-Scale Tracking (see Chapter 8)

This is best carried out in subdued lighting conditions with a monochrome input signal (preferably a test card), and it should be remembered that the video drive presets are for establishing the correct monochrome (grey) towards the high-lights of a picture, while the first-anode presets remove colour from the darker areas. Thus, if a primary hue (red, green or blue) predominates in the middle to highlights of the display the appropriate video drive preset should be reduced to give a neutral grey; and similarly, the appropriate first-anode preset should be adjusted at the darker end of the contrast scale.

Grey-scale tracking should be fairly correct on 405 lines after it has been adjusted properly on 625 lines; but if slight differences do exist between the two standards, it is generally possible to attain a compromise condition by adjusting the first-anode presets very slightly one way or the other from their settings originally established on the 625-line standard.

It is possible for grey-scale unbalance between standards to be caused by maladjustment of the 405- and 625-line drive presets. This is well worth remembering when adjusting these presets (if fitted). The scheme, referred to later, generally involves adjusting the presets to balance the boost voltage between the two standards.

Purity and Static Convergence (see Chapter 6)

Quite a lot has already been written about these subjects, but there are one or two practical points to bear in mind.

1. After adjusting for purity on a red raster, check on the green and blue rasters and, if necessary (though rarely likely) make compromise adjustments on these rasters with respect to the red one.
2. After completing purity check and, if necessary, adjust the red, green and blue radial and blue lateral convergence static magnets, concentrating, of course, on the centre of the screen.
3. If it is necessary to adjust these magnets, the purity should be rechecked, and if further optimising adjustments are made, the static convergence should again be checked and re-adjusted if necessary.
4. These procedures may have to be repeated to achieve the best possible purity and static convergence performance.
5. Unless substantial errors are present, however, indicating definite maladjustments, it should not be necessary to readjust the previously established correct dynamic convergence or to adjust the position of the scan coils on the tube neck.
6. If significant errors in dynamic convergence are indicated, then it is desirable to perform two sets of purity and static convergence adjustments—first a coarse set followed by a final, more accurate set. This is to minimise the effects of interaction between the static convergence and purity adjustments.

Full purity and convergence adjustments should not normally be made without reference to the maker's service manual, though, as the foregoing text has indicated, it is often possible to improve on the adjustments—once they have been made properly at the workshop, for instance—in the customer's home without reference to the manual. Indeed, this is often essential, because of the new position of the set in the customer's home, relative to its position when the full setting-up adjustments were performed.

Dynamic Convergence (see Chapter 6)

The exact procedure for dynamic convergence depends on the circuitry of the set (see Chapter 6), and when the set is completely out of adjustment it is desirable to have the manual at hand to assist with the adjustment; (eventually, of course, technicians working on one type of chassis in particular become highly conversant with the convergence procedures and can often make short-cut adjustments without the manual). For instance, it has been discovered that converging the Philips G6 chassis can be performed more speedily by first adjusting the red and green beams in the usual manner, but then switching off the green gun while the blue beam is being adjusted. This makes it easier to optimise the red and blue adjustments, while emphasising the effect of the blue adjustment.

Bad misconvergence can be seen on Test Card F, and two off-screen examples are shown at (a) and (b) in Plate 7, with the three guns active. Skilled technicians acquainted with the dynamic convergence circuit characteristics of specific sets can achieve almost perfect convergence using merely the background grid of Test Card F and the detail in the centre picture. However, for optimum results the type of display provided by a crosshatch generator is generally required. From first principles, adjustments are made first to converge the *vertical* centre lines of the display (as shown in Plate 7(c)) and then the *horizontal* lines. By repeating in turn the adjustments to the vertical and horizontal convergence controls the lines of the three colours on both axes first become parallel then coincident, giving the correct convergence condition with the least misconvergence at the edges and corners.

Perfect convergence over the entire screen area is rarely achieved in practice—not on the average domestic set, anyway—but the misconvergence in the corner areas of the screen should not be much greater than about 2·5 mm on a well designed and properly adjusted set.

Final adjustments along the lines just described can be carried out in a customer's home without reference to a manual, provided that the controls are only very slightly turned one way and then the other and immediately restored to their original positions if improvement is not apparent. It should be remembered that the workshop setting-up can quickly be destroyed by random, uncorrected adjustments to the convergence presets.

After completing this sort of convergence tailoring, it might be necessary to readjust- the purity magnet on the tube neck slightly for the best overall results.

At the time of writing, the BBC is radiating colour set installation information during the trade test transmissions of BBC 2, and some of the pictures which are displayed are extremely useful to the field technician. For the interest of readers not yet within range of a BBC 2 station, and overseas readers, Fig. 14.10 shows a couple of off-screen shots from these transmissions, taken by the author at a distance of about 100 miles from the nearest BBC 2 station. Figure 14.10(a) shows the nature of the dynamic convergence waveforms and (b) shows the effect of dynamic misconvergence, with the static convergence properly corrected.

OTHER ASPECTS OF ADJUSTMENT

Degaussing (see Chapter 6)

Thorough manual degaussing of the entire receiver is generally necessary before attempting to purify and converge—whether carrying out full-scale or final tailoring adjustments. Also, when the set is installed, large metal objects near to it must be degaussed. The auto-degaussing system is not sufficiently powerful to combat large internal (and, of course, external) spurious magnetic fields, so the field technician must carry a degaussing device around with him.

Horizontal Lock

Most sets have flywheel sync, and the best way to adjust the line lock control is first to remove the sync input lead (facilities are often available for this, or the set may have a 'push and twiddle' hold control knob—the pushing-in action removes the line sync) and then very carefully adjust the line lock control to the

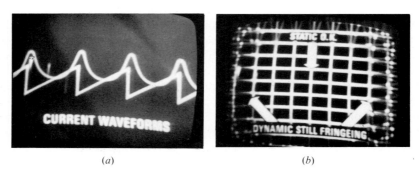

(a)   (b)

FIG. 14.10. *Off-screen photos of the BBC's trade demonstration film, showing convergence waveforms (a) and convergence display (b).*

approximately correct line frequency as shown by the picture just sliding horizontally. The picture will then lock solid on restoring the sync and remain locked on all transmissions. Many sets have either two presets and one line lock control, or two line lock controls, one for each standard, and it is necessary to adjust individually on each standard, depending on the design.

Pincushion Correction

The adjustment on the pincushion correction transductor is made first on the 625-line standard to make any residual bowing along the top border of the picture symmetrical. The adjustment is followed on the 405-line standard if necessary to secure a compromise correction between the two standards. It is noteworthy that electronic pincushion correction has been eliminated from the single-standard **BRC 3000** series by positioning the tube with blue gun at the bottom.

Horizontally Linearity

This control is usually adjusted in the same manner as that of monochrome sets; that is, making the adjustment on a test card to equalise the right- and left-hand sides of the display, relative to the circle in the centre. With many sets, the adjustment tends to affect mostly the right-hand side of the display.

Line Drive

This adjustment is more important in colour sets than monochrome, for it affects not only the nature of the horizontal scan, the e.h.t. voltage and the boost voltage, but also the focus potential. Moreover, it can change substantially between standards, and for this reason equalising 405-line and 625-line presets are often incorporated in dual-standard models.

Adjustment is best carried out on the 625-line standard by first tuning in a suitable transmission, blacking out the tube either by turning down the brightness control or turning off the three beam switches and then adjusting the 625-line drive preset for a specified e.h.t. voltage (this is usually 24 kV, but could be up to 25 kV on some models), after which the boost voltage should be noted (this is generally of the order of 1·25 kV when the drive adjustment is correct).

A 405-line transmission is then tuned in, the three beams muted as before, and the 405-line drive preset is adjusted for a boost voltage similar to that obtained on 625 lines. It may be necessary to obtain a compromise adjustment for the 405-line drive preset to give the best balance of boost voltages and e.h.t. voltages between the two standards. If the boost voltage tends to have an abnormally high swing on changing standard the grey-scale tracking could be affected, while if the swing is mostly on the e.h.t. voltage—the adjustment favouring boost voltage balance—then focus and static convergence unbalances could prove difficult to cure.

Third-Harmonic Tuning

It is rarely necessary to make any adjustments to the third-harmonic tuning of the line output transformer, for this is often fixed. However, if there are preset inductors or if the transformer assembly is replaced, the tuning should be made for maximum e.h.t. voltage—adjusting first the tuning on 625 lines and then that on 405 lines.

Incorrect third-harmonic tuning can incite line scan ringing symptoms, shown by vertical light and dark lines on the left of the picture.

Some single-standard sets using an e.h.t. tripler use fifth-harmonic tuning to give a flat-topped pulse and hence improved e.h.t. regulation.

E.H.T. Voltage Measurement

The e.h.t. voltage can be measured either with a 30 kV f.s.d. e.h.t. meter connected at the picture tube final anode—for example, a 20,000 Ω/V meter having an e.h.t. range extender (e.h.t probe) connected to it (see Chapter 11)—or with a high resistance, low reading voltmeter connected across a resistor which is generally included in the cathode circuit of the e.h.t. stabiliser tube. The actual voltage across the resistor corresponding to the specified e.h.t. voltage is given in the maker's manual, and this is 1·2 V when the resistor is 1,000 Ω and the stabiliser current is 1·2 mA.

Focus

Some dual-standard models have focus controls for 405 and 625 lines, and these are generally adjusted for the best focus performance, *on a bright picture*, after the adjustments referred to in the foregoing have been completed. However, it is worth noting that the setting of the focus controls is not particularly critical, and it is possible that the static convergence might suffer maladjustment before any serious changes in focus performance are observed when the controls are adjusted. This means, therefore, that any marked difference in static convergence between 405 and 625 lines should eventually lead to a check of the focus.

This may not apply to all models, because, as we have seen, some have switched balancing diodes in the line convergence circuits, which, in conjunction with preset (or fixed value) resistors, control the residual d.c. in the convergence coils.

It is also worth noting that static convergence errors between line standards can also be caused by incorrect width and/or e.h.t. adjustments on the affected standard.

Width

Dual-standard sets invariably have two width controls, one for each standard. These are adjusted in the usual monochrome-set manner, but in some colour models the controls are related to the e.h.t. voltage and line drive functions. Thus, if it has been necessary to adjust the controls substantially to balance the 405-line and 625-line widths, follow-up line drive and e.h.t. adjustments, as previously described, might also have to be made.

Picture Squaring and Shift

The picture is squared or levelled on the screen by rotating the scan-coil assembly on the tube neck in the ordinary way. The assembly is usually locked by one or two screws and it should be re-locked after adjustment, ensuring that it is pushed as far as possible along the neck against the flare, to avoid corner shadowing effects.

Because the convergence assembly is commonly secured to the rear of the scan assembly (often by means of a lockable bayonet fitting), adjustment to the scan assembly will possibly affect the positioning of the convergence assembly, and some convergence re-adjustment will be required.

The complete picture is shifted vertically and horizontally on the screen by shift controls which regulate the d.c. in the scan coils. On dual-standard models slight differences in picture position might be observed between standards. The adjustments in this case should be for the best compromise between 405 and 625 lines.

Field Timebase Adjustments

The controls associated with the field timebase operate in exactly the same manner as in monochrome sets and thus need no detailed reference.

COLOUR CIRCUIT ADJUSTMENTS
(ACCESSIBLE FOR THE FIELD TECHNICIAN)

Colour sets have an abundance of test points which are labelled TP and suffixed by a number and /or letter. Some of them are arranged for making adjustments in the colour circuits more simple, thereby making it possible for the field technician to undertake some of them in the customer's home.

Reference Generator Adjustment

This is one adjustment which can be handled relatively easily and, as an example, the procedure for a **RBM** receiver set is set out below.
1. Tune to a colour transmission, set the brightness and timebase controls for the best picture and turn the colour saturation control to its midway setting.
2. Over-ride the colour-killer circuit by connecting an 18 kΩ resistor to bias on the killer-controlled chroma amplifier stage (Fig. 14.11)—between test points TP.P2 and TP.P3.

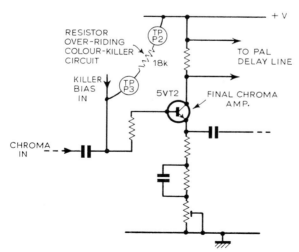

FIG. 14.11. *Method of disabling colour killer by connecting an 18k resistor between test points TP.P2 and TP.P3 (RBM).*

3. Adjust the reference generator trimmer capacitor 6TC1 to its midrange position (see Fig. 14.12).
4. Short-circuit the control potential input to the capacitor-diodes in the reference generator circuit, easily possible by connecting test points TP.E2 and TP.E3 together, as shown in Fig. 14.12.
5. Connect an oscilloscope to the output of the reference generator—to test point TP.E4 in the Rank set, as shown in Fig. 14.12.

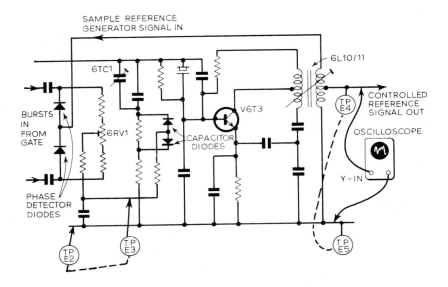

FIG. 14.12. *Reference generator (RBM), showing the various test points referred to in the text.*

6. Adjust the core of the output transformer 6L10/11 for maximum oscilloscope Y amplitude—e.g. maximum reference generator signal amplitude.
7. Transfer the oscilloscope Y input to any one of the colour-difference amplifier outputs.
8. Adjust the reference generator trimmer 6TC1 (Fig. 14.12) for 'zero beat' on the oscilloscope screen; that is, until the colour signals are just seen running freely on the screen (e.g. until the reference generator frequency is tuned as accurately as possible without a control potential).
9. Remove the short between TP.E2 and TP.E3.
10. Connect the oscilloscope (switched to d.c.) to TP.E3—the control potential input to the capacitor-diodes.
11. Mute the reference generator by shorting its output (by connecting test points TP.E4 and TP.E5 together).
12. Adjust the 'balance preset' 6RV1 (Fig. 14.12) for zero d.c.
13. Remove the short-circuit at generator output.

This concludes the reference generator setting-up, but to summarise the procedure for most sets, the fundamental actions consist simply of:

(*a*) over-riding the colour killer;

(*b*) adjusting the output coil for maximum output;

(*c*) adjusting at the colour-difference output for 'zero beat';

(*d*) muting the reference generator control potential;

(*e*) adjusting the manual tuning (preset) to colour-lock frequency (this can also be done by observing the colours on a picture and adjusting the trimmer until they slip as slowly as possible on the screen);

(*f*) adjusting balancing control for minimum d.c. control from phase detector.

A valve or transistor voltmeter (or a high resistance ordinary voltmeter—e.g. testmeter) can be used for most of these adjustments instead of an oscilloscope.

R — Y or V Chroma Channel Sync Adjustment

Some sets feature adjustments for the R — Y or V channel synchronising. It will be recalled that the ident pulses 'steer' the PAL switching in the R — Y or V channel so that the phase of the red or V detector automatically switches to coincide with that of the transmitted signal. If the detector switching is out of sync—e.g. out of phase—with the signal, then the reproduced colours will be the approximate complements of those of the real colours.

It is thus essential for the bistable circuit, which activates the red or V detector switching, to 'latch on' to the correct phase of the R — Y or V signal immediately the set responds to a colour transmission. When the sensitivity of this action is adjustable the basic procedure amounts to (*a*) optimising the ripple signal output by peaking the associated, high-*Q* tuned circuit, (*b*) turning down the amplitude of the ident signal, (*c*) interrupting the colour signal input until the red detector switches on colours, and (*d*) gradually increasing the amplitude of the ident signal by means of a preset resistor until the bistable is caused to 'steer' on to the correct switching phase, which is indicated by the picture colours becoming correct. This is the sync threshold point and when this is accurately established by the sync immediately locking on to the correct phase on channel changing, and when the input signal is restored after being interrupted, then the control should not be advanced any further as this could impair, rather than improve, the detector synchronisation.

It is, of course, out of the question to detail each and every adjustment of the various makes and models. All these are adequately covered in the service manuals. However, the adjustments referred to in the foregoing paragraphs are common to quite a large number of sets, as also are the interaction effects which their incorrect settings can incite.

FAULT LOCATING

In the majority of cases of trouble, the field technician will locate the source long before completing the whole string of installation and setting-up adjustments. In fact, the trouble may be readily revealed without the need to make any exploratory adjustments at all. For example, the classic symptom of *lack of colour, monochrome and sound all right* would point immediately to trouble in the chroma stages somewhere and one or two tests would soon show which unit or module is in need of replacement to restore normal working. Sometimes, until experience is gained or, indeed, simply to save time in field diagnosing, it may be desirable simply to substitute a suspect unit or module as a means of testing. Eight times out of ten it will be found that the suspect unit is, in fact, responsible for the trouble! This is certainly not a very scientific way of servicing, but it can save quite a bit of field time which is usually at a premium but often more so during the winter when one is extra busy. One can get to grips with the real servicing problems later, in the workshop, by clearing the faults on the returned units and modules if this is the nature of the procedure adopted by the service department. Alternatively, the faulty units and modules will have to be returned to the makers for servicing or replacement.

To sum up, then, field servicing includes (*a*) initial installation, (*b*) secondary setting-up procedure (though sometimes the full setting-up exercise might have to be run through), (*c*) locating the faulty section (Chapter 12), (*d*) replacing the responsible unit or module, and (*e*) finally making sure that the customer is completely happy with the repairs and/or adjustments.

INSTRUMENTS FOR FIELD SERVICING

The four main instruments required for field servicing are: the crosshatch and dot pattern generator with r.f. output; the external degausser; the high resistance testmeter (not less than 20,000 Ω/V ordinary testmeter or electronic testmeter— valve or transistor); and, preferably, an instrument providing a white reference.

More elaborate adjustments might be better performed with a small, portable oscilloscope, depending on the design of the sets which are serviced in the field. Some authorities claim that a colour-encoded pattern generator, incorporating crosshatch and dot facilities, can be advantageously employed in the field. This is possibly true, assuming that the field service technician has the degree of skill required to get the most from this expensive instrument. The trend would appear to be for this type of generator to be available in the service department and for the less costly non-encoded generators to be available for field servicing. A colour-encoded generator is certainly required somewhere in the servicing complex—perhaps more than one, depending on the magnitude of work undertaken.

Finally, mention must again be made of the signal strength meter. This is equally as important for colour television servicing as the multirange testmeter. Many poor reception troubles are caused by weak aerial signals, and only a meter capable of reading the signals at the end of the downlead direct can prove this quickly. It is thus essentially a field servicing instrument.

CHAPTER 15

Tuned Circuit Alignment

THE USE OF TRANSISTORS in the tuner and i.f. stages of the colour set tends to avoid progressive drift in tuned circuit alignment, not uncommon in fully valved sets, as the critical valves reduce in emission and change slightly in characteristics with age. Thus, once a transistorised i.f. strip is accurately aligned there will rarely be need to realign it completely.

However, a major servicing operation in the tuner or i.f. stages, involving the replacement of transistors and capacitors, will almost certainly modify the response characteristics, thereby calling, at least, for a run-through of the alignment procedure to restore optimum performance. The overall vision and sound responses are tailored essentially in the i.f. and intercarrier sound channels, while the tuner provides accurate channel tuning and delivers the correct i.f. signals to the first i.f. stage.

Most technicians now attempt to service v.h.f. tuners themselves, rather than returning them to the makers, which was fairly common practice in the early days. It is a bit different with u.h.f. and integrated v.h.f./u.h.f. units, however, for optimum performance of these depends on extremely critical alignment at frequencies up to 800 MHz, and it is generally best to return a misaligned or suspect unit to the maker for skilled attention, aided by a factory-based alignment jig. Indeed, this technique is not discouraged, particularly with colour sets, by the makers, who now supply reconditioned or new replacement units often on a return-of-post basis.

Some colour models are designed to a plug-in unit or 'module' pattern, so that in such sets trouble or misalignment in the i.f. stages might also be cured simply by replacing the faulty unit or module. Nevertheless, as colour sets age and technicians become more and more familiar with them, the urge to handle one's own repairs and re-alignment, even in a replaceable module or unit section, will probably lead to the makers supplying fewer reconditioned units.

Some of the symptoms of misalignment were discussed in Chapter 12 (see also Fig. 12.1) and, as explained in that chapter, a fair idea of how well the set is aligned can be discovered by operating the channel tuning control when the set is responding to a test transmission. At this point, however, it must be emphasised that sets incorporating automatic frequency correction (a.f.c.) will only yield possible fault information by display when the a.f.c. control is switched off or bias is disconnected from the tuner. This, of course, is because the a.f.c. tends to correct any tuning error. Moreover, symptoms of impaired alignment could be introduced by improper working of the a.f.c. circuits. For instance, if these circuits are incorrectly aligned or adjusted, the control bias may tend to detune the oscillator, rather than correct it.

RESPONSE CHARACTERISTICS

Since dual-standard colour sets are still in existence, we must consider dual-standard alignment, and to start with Fig. 15.1 shows typical i.f. response characteristics (*a*) for 625-line operation and (*b*) for 405-line operation. The various critical frequencies on both responses are indicated, and those marked with a *T* correspond essentially to tuned traps or rejectors in the circuit. It will be noticed that some of the indicated frequencies are common to both standards.

The circuit of a Mullard-designed dual-standard vision and sound i.f. channel is depicted in Fig. 15.2, and when this is correctly aligned the response characteristics shown in Fig. 15.1 result. It will be useful to identify the various tuned circuits in Fig. 15.2 and to see how they actually tailor the response curves.

TUNED CIRCUIT IDENTIFICATION

For the sake of simplicity of description, only the inductors associated with the tuning will be referred to below, the first of which is L201, the input to the vision

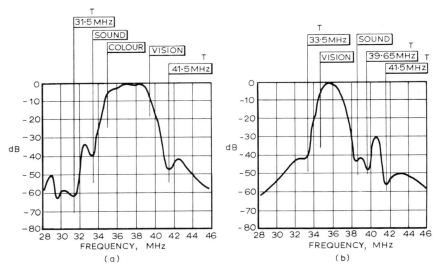

Fig. 15.1. *Vision response characteristics, upon which the important frequency points are marked. (a) 625 lines and (b) 405 lines.*

i.f. amplifier. L202 is a 33·5 MHz input trap, which is partly responsible for the −40 dB dip at that frequency in response (*b*) in Fig. 15.1, and for the setting of the sound carrier point in the 625-line response (*a*), also at −40 dB.

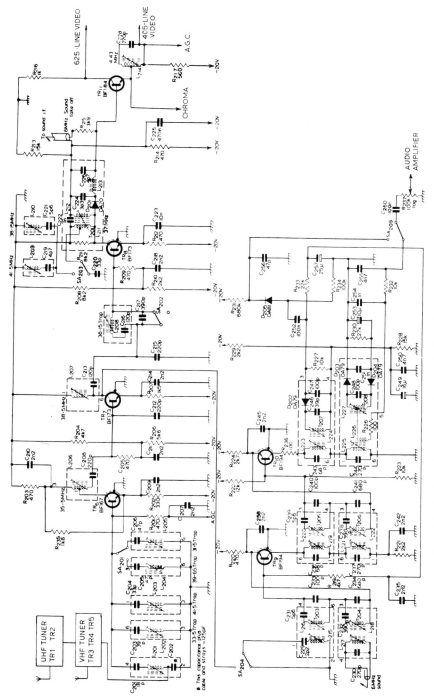

FIG. 15.2. *Dual-standard vision and sound i.f. channel (Mullard). The tuned circuits of this circuit are examined in the text, as also is the procedure for alignment.*

L203 and L209 are 41·5 MHz traps corresponding to the 625-line adjacent sound i.f. and Channel 1 sound i.f. L204 is a 39·65 MHz trap corresponding to the adjacent vision channel on the 405-line standard only (switched out on the 625-line standard). L205 is a 31·5 MHz input trap corresponding to the adjacent vision channel on the 625-line standard only (switched out on the 405-line standard). L206 is the first stage collector load and L207 is the second stage collector load. L208 and L210 are intermediate and output 38·15 MHz traps respectively, corresponding to the 405-line standard i.f. sound carrier and operating on the 405-line standard only. Transformer L215–L216 is also a 38·15 MHz 405-line sound trap, and it also acts as the 405-line sound take-off.

L211–L212–L213 assembly is the vision output transformer, while L214 is the chroma filter, tuned to 4·43 MHz approximately. Transformer L217–L218 is the intercarrier sound take-off, tuned to 6 MHz; and finally, assemblies L219–L220 and L223–L224 are 405-line sound i.f. transformers, with L221–L222 and L225–L226–L227 acting as the intercarrier i.f. transformer and ratio detector transformer respectively.

ALIGNMENT PROCEDURE

The various tuned circuits in Fig. 15.2 serve to give the response curves the shapes depicted in Fig. 15.1, but only when the circuits are very accurately tuned. The type of tuned circuits and, indeed, the design of the i.f. and intercarrier channels differ in detail from set to set (see Fig. 9.11). Even so, it will be a good exercise to examine the alignment procedure for the Mullard circuit.

The signal must first be applied either at the test point of a connected tuner or to the i.f. input via a circuit to simulate the i.f. output coil in the tuner; and secondly, some means must be provided for adjusting the gain of the amplifiers without a.g.c. action. This can be achieved by introducing a potentiometer into the a.g.c. circuit, across which is connected a suitable battery.

The signal should be monitored on the screen of an oscilloscope, the timebase of which is caused to sweep by a TV wobbulator over the required bandwidth (up to a maximum of 10 MHz, depending upon the section under alignment). The wobbulator signal is then applied to the input (see above) and its signal level is adjusted to give 2 V d.c. across the video diode load R215, when T201 is peaked to 37 MHz and with the circuit switched to the *625-line position*.

Traps L209 and L210 are then adjusted for maximum attenuation at 41·5 MHz and 38·3 MHz respectively, with the circuit switched to the *405-line position*. Traps L202, L203 and L205 are next tuned for maximum attenuation at 33·5 MHz, 41·5 MHz and 31·5 MHz respectively.

With the gain control (fitted as suggested above) adjusted to give an output 20 dB down on full gain, L201 is adjusted for the maximum level response centred on 37 MHz.

L206 and L207 are roughly adjusted to 35·5 MHz and 38·5 MHz respectively, and then carefully adjusted so as to set the vision carrier frequencies (34·65 MHz on 405 lines and 39·5 MHz on 625 lines) exactly 6 dB down on the peak response (see Fig. 15.1). Finally L201 is re-adjusted, if necessary, to remove any tilt at the centre of the passband. The last three mentioned operations should be repeated for optimum results, which concludes the 625-line vision i.f. alignment.

The circuit is then switched to the 405-line standard and L204 is adjusted for maximum attenuation at 39·65 MHz, L208 for maximum attenuation at 38·3 MHz and L215 to about 38·2 MHz. These final three adjustments put the various high-frequency end dips into the 405-line response curve, as shown in Fig. 15.1(*b*).

The sound circuits are next adjusted by connecting an indicating device (millivoltmeter or oscilloscope) to the audio output. L216, L219, L220, L223 and L224 are adjusted for maximum output at 38·15 MHz, with the lowest possible level of generator signal connected to the amplifier input.

The generator is then transferred to the junction of L213 and R215, via an 0·1 μF isolating capacitor, and L217, L218, L221, L222, L225 and L227 are adjusted for maximum output at exactly 6 MHz, keeping the level as low as possible to avoid limiting action.

SINGLE VISION DETECTOR

As already mentioned, various circuit configurations will be found in practice, many of which differ from that of Mullard. The Mullard circuit, in common with quite a number of others, adopts a single detector to provide luminance, chroma and intercarrier sound signals. This scheme works quite well provided that the sound carrier is sufficiently attenuated relative to the colour subcarrier. However, if the alignment is incorrect, giving inadequate sound carrier attenuation, intermodulation beats between the sound carrier and the colour subcarrier will almost certainly mar the pictures with 1·57 MHz pattern interference.

If the response curve deviates from the ideal, the response at the colour subcarrier frequency might either be exaggerated, where it will cause oversaturation and pattern effects, or attenuated, which results in desaturated colours and possibly colour noise symptoms—known as a 'rainbow' or 'confetti' effect.

Misalignment, causing the response at the 625-line sound frequency to be excessive (insufficiently attenuated), will not only aggravate pattern interference, as mentioned above, but it will tend to encourage intercarrier buzz symptoms, which also result from misalignment of the intercarrier sound channel or ratio detector transformer. Unbalance of the ratio detector diodes, or associated resistors, can yield a similar effect, and a variable resistor is often incorporated in one arm of the detector to secure optimum balance, as indicated by minimum intercarrier buzz.

A circuit using a separate detector for the intercarrier sound channel is given in Fig. 9.10 and one with a separate chroma detector in Fig. 9.9.

Mention should be made of the 4·43 MHz notch filter in the luminance signal path. This a substantial dip into the video response at the colour subcarrier frequency for the purpose of minimising the 4·43 MHz dot pattern. Its correct alignment is, therefore, very important.

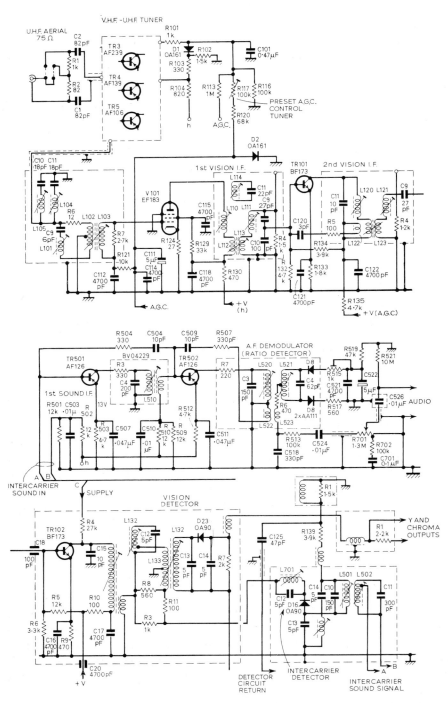

Fig. 15.3 *Vision i.f., intercarrier and detector circuits of the Körting ADO-1196 single-standard colour receiver.*

It is important, too, for the response characteristics to remain substantially as tailored by the various tuned circuits over the whole range of gain control, and this can sometimes present a problem—mostly of design—when transistors are employed in the i.f. channels. However, contemporary equipment, using the latest, low-capacitance transistors, and those specially designed for a.g.c., suffers considerably less from this effect. Nevertheless, it is just as well to check the response at various gain settings subsequent to realignment.

Needless to say, phase distortion in the vision i.f. channel must be at an absolute minimum, which is another reason why circuit realignment must be very carefully handled, for although somewhat incorrect alignment may give a fair response curve, some components of the composite colour-encoded vision signal may then tend to arrive at the vision detector before others (which is phase distortion, of course), and this will affect the colours, depending on the magnitude of phase error. In PAL-D sets, the main tendency is towards desaturation.

625-LINE MODELS

Single-standard, 625-line only models are far less complicated in their i.f. stages, and circuit realignment is that much easier to perform. It is a relatively simple matter to align the various tuned circuits for truly optimum 625-line only response characteristics, and there are fewer filters to get involved with.

The intercarrier sound channel is also simpler since it feeds just one detector— usually a ratio detector—and it is responsive to only one frequency—the 6 MHz intercarrier frequency.

These points are illustrated by the diagram in Fig. 15.3, which shows the vision i.f., intercarrier and detector circuits of the German single-standard Körting receiver, Model ADO-1196.

An unusual feature is the valved first vision i.f. amplifier stage. This is gain-controlled and picks up its signals from the v.h.f./u.h.f. tuner. It is bandpass-coupled to the second and third stages, which are transistorised. The vision detector feeds luminance and chroma signals to the monochrome and colour sections of the set, while an extra detector is used for the intercarrier sound signal (6 MHz), this signal is then amplified and detected by two transistor stages and a ratio detector (two diodes), shown at the top of the diagram.

CHROMA CHANNEL ALIGNMENT

It has already been mentioned that an advantage of a separate intercarrier detector is that a 33·5 MHz rejector can be used in the circuit of the vision detector to help to defeat the beat between the sound intercarrier signal and the chroma signal. Such a rejector cannot be used for great attenuation in a detector delivering intercarrier signal, of course, since the added attenuation would impair the production of the intercarrier signal, which itself relies on the beat between the sound and vision carriers.

It will be recalled that the chroma subcarrier is placed a little down one side of the i.f. response while the vision carrier is placed about 6dB down the opposite side (for normal single-sideband detection). In order to reduce quadrature distortion in the chroma channel, it is not uncommon to find that the chroma channel alignment yields a response that is not perfectly flat over 1 MHz either

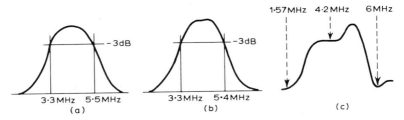

FIG. 15.4. *Response curves of the chroma channel in single-standard receivers. The alignment exercise may call for the tuned circuits to be adjusted first to obtain curve (a) and then tailored to achieve curve (c) or curve (b). Curve (c) represents the chroma response of the BRC sets.*

side of the chroma subcarrier frequency. This combats the unequal amplification of the chroma sidebands.

Figure 15.4 shows chroma response curves (a) for symmetrical response, (b) for slight asymmetry and (c) for the shape which may be recommended for the least quadrature distortion. The final shape (c) would be obtained by careful

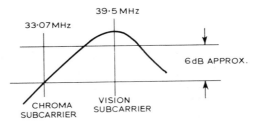

FIG. 15.5. *Sets with a chroma i.f. channel might have the alignment adjusted to give the chroma i.f. response raised amplitude at the vision carrier frequency (39·5 MHz). The response curve shown here is that relating to the Bang and Olufsen 3000 series single-standard models.*

adjustment to the chroma bandpass tuned circuits, and for this an oscilloscope/wobbulator combination is essential.

When the chroma detector is fed from its own i.f. channel, the response tailoring can be achieved in the i.f. alignment, and Fig. 15.5 shows the response recommended for the Bang and Olufsen 3000 series single-standard models. Here the amplitude of the 39·5 MHz signal (the vision carrier) is raised in order to reduce quadrature distortion of the chroma signal.

CHAPTER 16

Faulty Picture Tube Symptoms

SINCE a shadowmask picture tube features three separate electron guns, one for each primary colour, it is possible for a fault to develop in one without the other two being affected. Some of the faults that trouble the guns have their parallel in monochrome tubes, which are mainly low emission, electrical leakage between electrodes, inter-electrode shorts and open-circuit electrodes. There are other faults which affect all three guns simultaneously, some of which are common only to colour tubes. These include open-circuit heater (this affects the three guns because the heater is shared equally by all of them); worn or burnt phosphors; and buckled and/or misaligned shadowmask.

FAILURE OF ONE GUN

If one gun goes open-circuit completely—i.e. fails to deliver beam electrons—then the primary colour related to it will be missing from the picture. When the tube is handling a monochrome transmission, or displaying a colour transmission in monochrome, this trouble is immediately apparent by the picture being predominant in a complementary hue—magenta, cyan or yellow. See Plate 5(a), (b) and (c). Indeed, it is always desirable to have in mind the nature of the monochrome display produced by failure of one gun, as shown in Table 16.1.

Table 16.1

Faulty Gun	Predominant Complementary Hue
Blue	Yellow
Red	Cyan
Green	Magenta

The display will also be biased towards the complementary hue should the beam current of the gun of the associated primary colour be below that required to produce the correct monochrome white on the screen. Thus, a low emission blue gun, for example, would tend to turn the monochrome display towards yellow.

HIGH BEAM CURRENT

Conversely, a monochrome display biased towards one of the primary colours would suggest that the associated electron gun is delivering a larger beam current than required for a correct monochrome white display. This could mean in the case of a predominantly blue display, for instance, that both the red and green guns are low in emission, but this symptom would be more likely to be caused by maladjustment of the grey-scale tracking; i.e. the adjustment favouring the current of the blue beam over a part, or the whole range, of the tube drive and/or brightness control.

CHARACTERISTIC CHANGE

However, it should be borne in mind that the grey-scale tracking can drift because of a change in the characteristics of one or more of the guns. Normally, of course, sufficient adjustment range is available on the cathode drive and the first-anode presets to compensate for a reasonable characteristic change of this kind; but if the change is excessive the presets might be out-ranged and difficulty may be experienced in restoring a fair overall adjustment of grey-scale tracking.

LOW BEAM CURRENT

Should the symptom point to trouble in one of the guns, then the display provided by that gun alone is best observed on the screen. This is easily achieved, of course, by switching off the other two beams. Let us suppose, for instance, that a monochrome picture has a predominantly yellow background, suggesting a faulty blue gun, then the plan would be to switch off (or back-bias) the red and green guns, leaving the display in blue only.

This display should normally change in intensity over the whole range of the brightness control, be free from the effects of saturation or limiting at high beam currents and, of course, disappear from the screen when the brightness control is fully retarded. In other words, the one gun should behave as any normal monochrome gun.

Since the symptom suggests low beam current of the blue gun, subjective tests will almost certainly reveal low blue brightness over all or part of the range of the brightness control. Low emission of this gun or, indeed, any gun could also incite the typical monochrome tube effect of limiting, whereby, instead of the display continuing to increase in brightness as the control is turned towards the top end of its range, the bright parts take on a glistening, silvery appearance. This is caused by the gun cathode failing to provide all the electrons required by the beam for a display of maximum intensity.

A worn tube could give this symptom on one or more of its displays, but since the red gun usually has to provide a greater beam current than the green and blue guns to maintain correct monochrome balance—owing to the lower light output of the red phosphors compared with the green and blue ones (the red gun, in fact usually requires a beam current 10 per cent higher than each of the other two guns to produce white)—this gun not uncommonly tends to exhibit low emission symptoms before the other two.

BOOSTING

While in the past—when early monochrome tubes were expensive—it was the practice of service technicians to boost a worn tube by over-running its heater in an endeavour to increase its useful life, this is not generally possible with colour tubes when one gun drops badly in emission relative to the others because the heater is common to the three cathodes.

ELECTRODE OPEN-CIRCUITS

One way in which a gun can fail completely relative to the other two is by an open-circuit developing between the pin and the electrode of the cathode, control grid or first anode. The second anode (focusing electrode) and the third and fourth anodes (composing the final anodes to which the e.h.t. voltage is applied) are common to the three guns, so a fault in these will usually affect the three guns together.

If the cathode or first anode of one gun goes open-circuit, for instance, the beam current of that gun will collapse, leaving a monochrome picture predominated by the faulty gun's complementary hue, as already explained. If the control grid goes open-circuit, on the other hand, the beam current of the affected gun will rise towards maximum and cease to be influenced by the brightness control.

An open-circuit developing in the focus electrode or final anode system will cause the tube to fail on all three guns; and this, of course, also applies to an open-circuit heater. Open-circuit electrode faults, however, are comparatively rare, as they are in monochrome tubes.

INTER-ELECTRODE SHORTS AND LEAKS

Modern colour picture tubes are highly engineered to strict standards ensuring long, trouble-free service, and for this reason the faults now being described are not very frequently encountered. Nevertheless, the symptoms they yield must be fully understood, and with this in mind we will now consider the symptoms produced by short-circuits and electrical leakages between the electrodes of the separate guns.

These symptoms are well known so far as they affect monochrome picture tubes and, looking at the three guns of a colour tube in terms of three independent tubes, the symptoms resulting from inter-electrode shorts and leaks correspond closely to those resulting from similar faults in the solitary gun of a monochrome tube.

Heater-to-Cathode Short

This fault can give rise to a diversity of symptoms, depending on the design of the receiver. The tube heater in many receivers is energised either from a separate secondary winding on the mains transformer, or from a tapping on the primary winding, and acts in relation to the tapping as an auto-transformer. If the winding is connected to chassis (e.g. 'earthed'), a heater-to-cathode short will remove the positive potential from the cathode and thus cause the control grid of the affected gun to assume a substantial positive potential with respect to the cathode, as shown in Fig. 16.1.

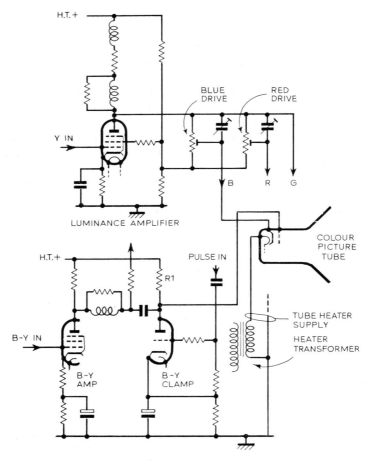

FIG. 16.1. *Colour-difference picture tube drive, showing the luminance output stage and one of the colour-difference amplifiers with its clamp.*

This shows a typical circuit of valved luminance and colour-difference amplifiers complete with the clamp triode circuit on the grid. The grid is rendered positive by R1 on the triode anode being returned to h.t.+ line, while the conduction of the triode determines the black-level reference. It will be recalled that this triode is 'gated' by pulses fed in from the line output stage.

In the type of circuit shown, brightness is controlled by varying the conductance of the luminance amplifier, and hence the potential at the valve anode, by the control regulating the luminance amplifier bias. Thus, with the cathode down to chassis this action will be destroyed, and the gun will continue to supply beam current, of a value dependent on the grid voltage reflected from the colour-difference amplifier anode, at all settings of the brightness control. Moreover, the short will be reflected on to the other two guns, via the electrical continuity in the cathode feed circuits, and the Y signal will be bypassed to chassis. The symptom will thus be of uncontrollable brightness with possibly, depending on design, any background display resulting only from the colour-difference signals. In some cases a vestige of Y signal may get through to the fault-free guns depending on which gun is at fault and on the setting of the drive presets.

For example, if the blue gun in Fig. 16.1 is at fault and the blue drive preset is about 50 per cent advanced (determined initially by the grey-scale tracking, of course), then the heater-to-cathode short will fail completely to bypass the Y signal and some will get through to the red and green cathodes. However, the grey-scale tracking will be very badly disturbed, to say the least! On the other hand, if the green gun is the faulty one, a heater-to-cathode short will bypass all the Y signal because the circuit in Fig. 16.1 shows the green cathode feed taken direct from the luminance amplifier valve anode, instead of via a preset or 'hold-off' resistor.

If the supply feeding the tube heater is isolated from chassis, then a heater-to-cathode short on one of the guns will fail to affect the biasing of any of them. The maximum effect will be attenuation of the higher-frequency components of the Y signal, which is mainly caused by capacitance shunting of the luminance amplifier by the heater transformer. The Y definition will thus be impaired and some degree of mis-registration between the luminance and colour elements of a picture may result due to the slowing-down effect that the reflected shunt capacitance might have on the Y signal.

It will be recalled that a Y delay line is used earlier in the luminance channel to 'slow down' the wideband Y signal so that it arrives at the tube cathodes at the same time as the narrower band colour-difference signals arrive at the grids. Extra capacitance loading in the Y channel could thus disturb this critical timing.

Heater-to-Cathode Leakage

This fault will give symptoms similar to those of a complete short, though in a somewhat less severe manner. For instance, the Y signal will never be completely bypassed to chassis (assuming that the heater circuit is 'earthed'—

connected to chassis) and the brightness control action will not be totally disabled, depending on the magnitude of the leakage.

Nevertheless, the grey-scale tracking will be substantially impaired, and it will almost certainly be discovered—when the affected gun is run in isolation—that the raster is not completely extinguished when the brightness control is fully retarded.

This fault in moderation can sometimes lead to the conclusion that the grey-scale tracking adjustments are in error, and one may well attempt to remove the symptoms by re-adjusting the tracking presets. Indeed, this can sometimes result in fair grey-scale tracking, even when one of the guns possesses a heater-to-cathode leak (a very small one), but the tracking performance under this condition will never be as good as it should be.

This does go to show, however, that a fault condition can be masked, albeit inadvertently, by re-adjustment of the tracking presets but this procedure is, of course, never to be recommended, for once an electrical leakage develops it rarely stabilises to a constant value. This means that the re-adjusted grey-scale could drift badly as the tube rises in temperature and the leakage value changes.

Heater-to-cathode shorts and leakages can also introduce the symptoms of hum, as explained in Chapter 12.

Grid-to-Cathode Leakage

Although it is possible to get a total short-circuit between the grid and cathode of one of the guns, this is singularly unusual; the most common fault in this respect is electrical leakage between the two electrodes. Again, the prime effect is that of incorrect gun biasing, but this time the Y signal is not significantly affected or attenuated. If the leakage is pretty small, the symptom on the affected gun will be of misplaced beam current cut-off. If the beam does, in fact, extinguish, this will happen after the other two beams derived from the fault-free guns have extinguished. This again seriously affects the grey-scale tracking.

If, say, the red gun is at fault, then the picture will possess a predominantly red tint over a part or all of the brightness control range. This is because the leakage incites a greater beam current from the faulty gun because of the resulting fall in cathode-to-grid bias.

Grid Emission

Similar symptoms to those just described, though possibly to a smaller degree, occur as the result of the grid of one (or more) of the guns acting as a cathode. This might be brought about by small particles of cathode material coming into contact with the grid and the grid then producing electrons, as well as the cathode, as its temperature rises.

The quantity of electrons so produced is much smaller than produced properly by the cathode, but the fact that the grid is acting as an emitter of electrons, although in a very small way, means that it rises a little positively with respect to the cathode of the same gun. This tends to unbalance the grey-scale tracking, of course, by increasing the bias a little above the original preset value.

TEMPERATURE EFFECTS

As the above symptom tends to come on progressively as the tube warms up, the grey-scale tracking might well be perfectly normal for a little while after the set is switched on, but it then drifts significantly afterwards. If the picture tends to take on, say, a blue tint after the set has been running for 30 minutes or so, then it is possible that the blue gun is suffering from grid emission. It is noteworthy that a similar symptom can be caused by drift in the colour-difference stages— active device (such as valve or transistor) or passive resistor or capacitor.

Some of the other faults described can also be influenced by temperature increase of the tube. The emission, for instance, of one gun may be relatively low to start with, rising gradually to normal as the tube warms up (this has a strong parallel in monochrome tubes as is well known). Under normal conditions of grey-scale adjustment, therefore, the picture would exhibit the complementary hue of the affected gun when first switched on (e.g. showing a cyan tint with the red gun low emission), and would gradually become normal as the tube warmed up and the emission rose towards normal by yielding the correct ratio of beam currents for white.

If the grey-scale is re-adjusted *before* the emission of the bad gun rises to its normal value, the effect with increasing emission of the gun will be that the display will become progressively tinted with the *primary hue* associated with the defective gun. Thus, if the green gun is at fault, the display will assume a greenish tint as the emission gradually rises.

The degree of heater-to-cathode leakage can change with rising temperature of the tube and, similarly, almost any tube defect can change with respect to temperature.

For accurate grey-scale tracking the characteristics of the three guns have to be accurately matched by the drive and first-anode presets (see Chapter 8). This means, therefore, that any unbalanced change in the characteristics of the guns will affect the tracking performance, demanding complete re-adjustment to restore the original performance. This might happen normally during the long life of the shadowmask tube and, provided that the drift is not short-term, it will not matter, but if it is, then the tube would have to be replaced to avoid the viewer having to employ the services of a resident technician!

DRIVE CIRCUITS

Most of the effects discussed are common to all receivers, but the nature of the symptoms can differ between receivers depending on the precise design of the tube cathode and grid circuits. The circuit in Fig. 16.1 adopts the very popular colour-difference drive system, whereby the Y signal is fed to the cathodes together, via the drive presets for grey-scale tracking, while the red, green and blue colour-difference signals are fed to the grids for corresponding colour identification. In this system the tube itself performs the matrixing action, yielding the primary-colour signals between the grids and cathodes of the three guns.

Some sets employ a separate matrixing circuit which receives the Y signals and the colour-difference signals and delivers true primary-colour signals which are fed to the tube grids or cathodes, and integrated circuits are now being used here. If fed to the grids, the signal voltages have to be higher than if fed to the cathodes, and this also influences the design of the circuits.

The usual scheme when primary-colour signals proper are used, therefore, is to feed them to the cathodes, and Fig. 16.2 shows the basic circuit of one of the primary-colour channels in the all-transistor BRC chassis which employs this arrangement. Here Y signal delivered at the emitter of the luminance amplifier

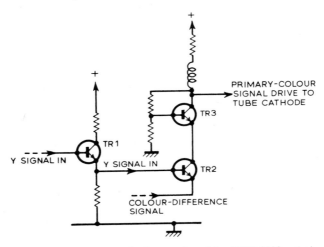

FIG. 16.2. *Basic circuit of primary colour drive (BRC 2000 series).*

TR1 is fed to the base of the matrix transistor TR2, which also receives the appropriate colour-difference signal at its emitter. This transistor subtracts the Y signal from the colour-difference signal, leaving only the real primary-colour signal in the collector circuit. This is then fed to the emitter of the primary-colour output transistor TR3 (arranged in a common-base circuit), thereby delivering the amplified primary-colour signal at its collector. This signal is then fed to the appropriate cathode of the picture tube.

Although this circuit is described in far greater detail in Chapter 8, it is referred to again to illustrate how tube faults give symptoms related to the nature of their drive circuits. For example, a heater-to-cathode short in one of the guns using primary-colour signal drive would fail to have much effect on the signals applied to the other two guns. Thus, the two normal guns would continue to supply correct beam currents while the faulty gun would be severely under-biased owing to the lack of cathode potential, assuming, of course, that the tube heater is at chassis potential. If the heater is isolated by a transformer winding, then the definition of the display produced by the faulty gun would almost certainly be below standard and thus out of balance with those produced by the fault-free guns.

OTHER TUBE FAULTS

As in monochrome tubes, troubles can arise due to misaligned electron guns, but with colour tubes this fault is generally more serious because it can make it impossible to secure correct purity. Similar troubles can also be caused by a buckled shadowmask*, but fortunately faults of this kind are few and far between.

It is just possible for a colour picture tube to suffer mechanical damage in transit without imploding, in which case difficulty will certainly be experienced in obtaining correct purity. If colour tubes are eventually rebuilt, then some degree of mechanical misalignment and/or shadowmask buckling or warping might be experienced, as gun displacement is infrequently experienced in rebuilt monochrome tubes.

Full information for obtaining optimum purity is given in Chapter 6, and this procedure should certainly be carefully followed, along with adequate de-gaussing activities, before a tube is suspected of being mechanically unsound.

PURITY CHECK

Another check for purity, which will reveal the goodness of the tube mechanically, is to observe the landing of the electron beam triads relative to the screen phosphor dot triads by means of a microscope with a magnification of, at least, 40 times. Mullard suggest that the phosphor dots may be illuminated by an external light source falling on to the screen at an angle of approximately 10 to 15 degrees. With all three guns switched on it is then possible to observe the landing positions of the three electron beams relative to the corresponding phosphor dots at the centre of the screen.

Correct purity is usually shown by the triad composed of the three beams landing in the geometric centres of the dots comprising the corresponding red, green and blue phosphor triads, but with some tubes—the Mullard 25 in. A63-11X, for example—optimum overall purity is registered when the three scan spots are slightly inset from the phosphor dots in each triad.

The Beam Landing Pattern Scope by Philips, shown in Fig. 11.6, is ideal for these observations. It can also be used as an alternative means of adjusting for purity by orientating the purity magnet on the tube neck until the conditions just explained are achieved. Plate 3 shows the illuminated dots under a microscope.

* A symptom recently investigated was that of a 'ping' in the tube followed by severe impurity of display a few minutes after warm-up. This was due to the shadowmask buckling with increasing temperature.

CIRCUIT FAULTS GIVING TUBE DEFECT SYMPTOMS

It is not proposed in this section to detail the various circuit faults that can disturb the correct operation of the colour picture tube, for these are adequately covered in Chapters 12 and 13. However, it is important to bear in mind that circuit faults can produce symptoms not unlike those associated with some of the tube defects previously considered in this chapter. One must thus decide early during the course of diagnosing for these sorts of troubles whether the fault lies in the tube itself or in its controlling circuits.

Checking for Tube Leakage

A definite short-circuit between the electrodes of the guns can easily be determined by connecting an ohm-meter across the pins corresponding to the suspect electrodes on the tube base when the set is switched off. Slight leakage, however, cannot always be detected in this way because the resistance it represents may be very high, especially when the tube is cold and inactive. There is a way, though, that quickly shows whether the tube or its circuits are at fault. Normally, the tube is biased by its circuits, and this bias should remain exactly the same whether the tube is connected or not. If the tube is faulty, then the bias will show a change when it is connected, and it is upon this basis that the tube can be checked for leakage.

It is best to mute the line output stage and disconnect the focus potential before applying the test, for it requires the removal of the tube base when the set is running. The focus potential is applied to the second anode, usually via pin 9, so this pin should be disconnected from the 5 kV source, either at the tube base or, if more convenient, at the second anode source itself.

The line output stage can generally be muted by lifting the screen grid feed from the line output valve, but a resistor should then be applied across the h.t. supply to simulate the total load normally applied by this valve, remembering that the anode current as well as the screen grid current will fall when the h.t. supply to the screen grid is removed. A 2 kΩ resistor across a line of 200 V will burn up 100 mA of h.t. current, a fairy average loading, but to prevent the resistor burning out or overheating it should be rated at around 100 W.

If the e.h.t. supply is derived from a generator which is isolated from the line output stage—which is unlikely—this should also be muted to prevent the tube from receiving final anode potential. It is also necessary to provide for heater chain continuity (with a resistor to simulate the tube heater) if the tube heater is included in the chain. However, as already mentioned, the tube heater is usually energised from a transformer winding.

The plan, then, is simply to measure the voltage across the cathode and grid tube base tags corresponding to the suspect gun—with the set switched on, of course—note the reading, and observe whether this changes when the base is removed from the tube. If it does, it can be concluded that some leakage in the

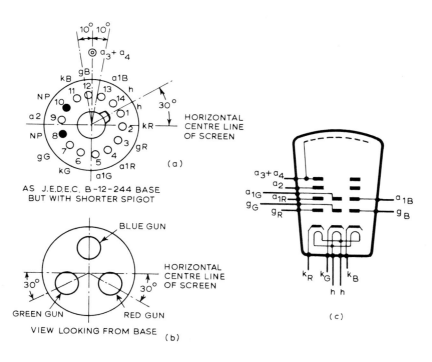

FIG. 16.3. *Colour picture tube (Mullard A49-11X), showing the base connections (a), the gun positions (b) and the symbol (c).*

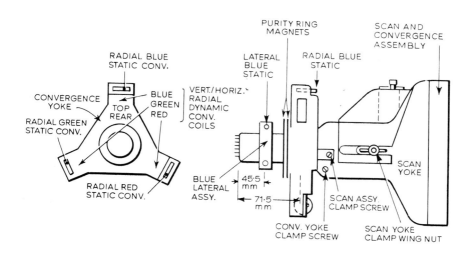

FIG. 16.4. *Assembly of the various components on the colour tube neck.*

gun is responsible for changing the circuit-derived biasing conditions. A high resistance voltmeter must be used for this check.

Artifices for clearing inter-electrode leakages in monochrome tubes have been adopted over the years by service technicians with varying degrees of success. Leakages between grid and cathode, and heater and cathode can sometimes be cleared by 'flashing' the pulse potential, which is present at the anode of the line

FIG. 16.5. *Neck components in detail. Left to right, scan yoke, convergence yoke, purity ring magnets and blue lateral assembly (RBM) (courtesy, The Rank Organisation).*

output valve, across the leaking electrodes. It is possible that this may also work with colour tubes, but at present there is no conclusive evidence available concerning this.

Since colour tubes are so expensive, it would appear worthwhile to try almost any scheme to clear a gun fault, provided the tube is totally useless in the fault condition; but tubes which are, at least, giving a fair colour picture, even though not perfectly sound, are best left alone, for any corrective procedure is likely to cause the tube to fail completely. This applies, too, to 'boosting' by over-running the heater, which was practised extensively in the days when monochrome tubes were relatively expensive.

CATHODE DRIVE CONSIDERATIONS

In Fig. 16.1 it is shown that the green cathode receives maximum video drive, since this is connected direct to the anode of the luminance amplifier valve, while the red and blue cathodes are fed via the drive presets. On the other hand, some other sets feed the red cathode direct from the anode and the green and blue cathodes through drive presets, and it is not uncommon to find changeover links in the cathode feeds from one of the presets and the anode. These allow either the red or green cathode to be driven direct from the anode, as required by the colour tube employed.

The reason behind this is that some tubes call for maximum red video drive and others for maximum green video drive. Unless some provision is incorporated for matching the maximum drive to the tube, a tube requiring, say, a maximum green drive will lack green in the display (giving magenta) if the red cathode is fed direct from the anode while the green cathode is fed through a drive preset. This consideration should also be borne in mind when replacing a colour tube.

It has already been mentioned that most tubes require a greater red beam current than green or blue beam current to give a white display. The red gun is thus generally driven harder than either of the other two. This means that its efficiency is likely to diminish before that of the green or blue gun, giving a cyan-tinted picture.

TUBE REPLACEMENT

There is little point in repeating information given in the manufacturers' service manuals on tube replacement, for each model calls for its own specialised procedure in this respect, and it is important to follow the step-by-step instructions when replacing a tube costing up to £100. Nevertheless, one or two hints and tips relating to the subject could be useful.

1. A colour picture tube is much heavier than its monochrome counterpart, some weighing in excess of 15 kg.
2. Only the correct replacement or substitute tube recommended by the maker of the set should be used. A substitute tube may, in fact, call for modifications to the controlling circuits, but these are detailed by the manufacturer specifying the substitute.
3. Always use protective clothing, or at least goggles, to protect the eyes from flying glass in the event of an implosion when removing an old tube from the set and installing a replacement.
4. Check on the positions occupied by the neck components before taking them from the old tube so that they can be correctly assembled on the neck of the replacement tube.
5. Locate the blue gun on both the old and replacement tubes and arrange that this position is occupied by the blue gun of the replacement tube (see Fig. 16.3).
6. Ensure that the metal rimband of the replacement tube is connected either direct to the chassis of the set or via a suitable leakage path as provided by the maker.
7. Ensure that the external conductive coating on the tube flare, acting as one electrode of the e.h.t. smoothing capacitance, is adequately connected to the chassis of the set.

315

8. Ensure that the auto-degaussing system and the magnetic shield are correctly positioned on the replacement tube.

The base connections, gun positions and the symbol of the Mullard A49-IIX 25 in. tube are given in Fig. 16.3. Figure 16.4 demonstrates the positions of the components on the tube neck, while Fig. 16.5 shows the same components in considerable detail.

CHAPTER 17

Receiver Design Trends

SINCE the launching of the first edition of this book a number of new developments have occurred in receiver design. The basic PAL system of transmission and reception remains unchanged, of course, so the earlier chapters are as equally applicable today as they were when the book was first written. However, to keep the reader in alignment with current trends it was considered desirable to include this new chapter—not so much to explore each new development in detail, but more as a kind of development summary.

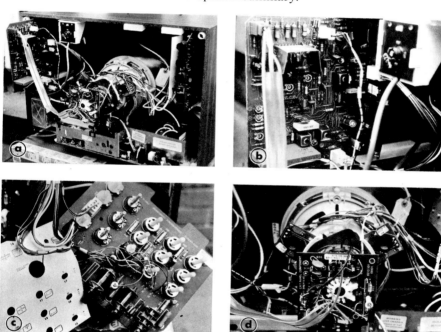

FIG. 17.1. *Internal views of the Pye 713 chassis, showing at (a) the printed circuit board mode of construction with the chroma board on the left partly removed, at (b) details of the chroma board which contains four i.c.s and performs all luminance and chroma signal processing to yield RGB drives for the picture tube cathodes and, apart from three RGB output transistors, almost all the active circuitry is incorporated within the i.c.s, at (c) the convergence board which contains the controls for dynamic convergence and grey scale adjustment, the board being removable from the cabinet and securable to brackets so allowing adjustments to be made from the front of the receiver, and at (d) details of the picture tube connector board.*

Readers interested in in-depth study of all the various circuit sections of the new models as they come along are referred to the author's volumes of *Newnes' Colour TV Servicing Manual* (by the same publisher). At the time of writing Volumes 1 and 2 have been prepared dealing with receivers of up to about 1974 vintage. All the circuit sections are examined in detail, and the books also include detailed portions on servicing and adjustments, with oscillograms. The plan is for those volumes to complement this book, thereby making conveniently available to the service technician, student and enthusiast a complete and expanding story of colour television.

MODULES AND I.C.S

Recent developments have resulted in the circuit complex appearing less involved than hitherto. This has ensued from the use of plug-in printed circuit boards and the so-called modular construction, with increasing application of integrated circuits and other solidstate devices. Some idea of the plug-in printed circuit board idea can be gleaned from the photos in Fig 17.1 of the inside of a receiver using the Pye 713 chassis. Cable links between 'mother' socket assemblies secure the required circuit communications between the various departments. The boards can be easily removed for replacement or repair, allowing the field service technician to correct a complex fault *in situ* merely by locating the faulty section and replacing. The faulty board can then either be returned to the manufacturer for repair or (as now more often being considered better and less frustrating) repaired at base by the bench technician.

The Pye 713 chassis is one which well summarises the design trends. Another is the British Radio Corporation's 8000 chassis (taking in the 8500 series).

PI TUBE

The shadowmask tube still predominates the scene, and Sony has developed a larger 114 degree version of the Trinitron tube (page 27). Other display tube developments are appearing, a notable one being the 90 degree 'Precision in Line' (PI) tube introduced to the UK by Thorn and developed in the US by RCA. Like the Trinitron, the PI tube has the three electron beams in horizontal plane,

(a)   (b)

FIG. 17.2. *The slotted apertures (a) and the corresponding phosphor stripes of the PI display tube.*

which neatly (as with the Trinitron) confines convergence adjustments essentially to mere horizontal shift.

The display end consists of a mask carrying a pattern of slit-shaped apertures and corresponding green, red and blue phosphor stripes, but not of full length like the Trinitron (Fig. 17.2). This technique makes possible the use of the slight spherical curvature effect characteristic of the shadowmask screen. With the full-length stripes of the Trinitron the screen can be curved only in the horizontal plane.

Each stripe is 0·0108 in. wide, giving a one-colour stripe repeat at 0·0324 in. pitch, this ensuring good horizontal resolution even with small-sized tubes (16 in. being the smallest), and better with 18- and 20-in. versions.

Colour purity is adjusted by magnetic fields in the conventional manner, but with the requirement being confined to the horizontal only. Static convergence adjustment is by a set of four ring magnets. One pair yields a four-pole field with zero field at the axis along which the red beam passes. This pair, therefore, adjusts the blue and green beams either towards the axis and hence red beam or the blue beam downwards and the green beam upwards. The second pair yields a six-pole field, again with zero field at the axis and this adjusts the blue and green beams both in one direction horizontally or both in one direction vertically, depending on the adjustment sequence of the ring pairs.

The scanning yoke, called 'Precision Static Toroid' (PST), is designed specially to minimise dynamic misconvergence, and is integral to the tube when purchased, thereby always ensuring optimum alignment and hence removing the need for dynamic convergence adjustments. This arises because in correct alignment the strength of the deflecting field applied to each beam corresponds to the requirement for the correction of dynamic misconvergence. In other words, the strength of the fields differs between each beam.

Other recent in-line gun, vertical stripe screen tubes include the 26 in. 110 degree 'quick vision' (reducing warm-up from 15s to 5s) Philips 20 AX—in the UK by Mullard—and a range (14, 16, 18, 20 and 22 in.) by Toshiba with 90 and 110 degree deflection angles, all with simplified dynamic convergence.

COLOURING CIRCUITS

Integrated circuits are replacing large areas of discrete components. Such devices are now found in the colouring circuits as well as in other departments. The Pye 713 chassis, for example, incorporates no fewer than seven main i.c.s.

A simplified block diagram of the i.c.s used in the video circuits of the Pye 713 is given in Fig. 17.3. This model, which, incidentally, is fully covered in Volume 2 of *Newnes' Colour TV Servicing Manual*, adopts primary colour or RGB drive for the picture tube. The TBA560 accepts and amplifies the luminance signal applied to pin 3 from the luminance delay line L292 and communicates the signal to the primary colour matrix i.c. TBA530. The colour-difference signals are also fed to this i.c. from the main decoder i.c. TBA990, the matrix then yielding the primary colour signals—i.e., R derives from $(R - Y) + Y$

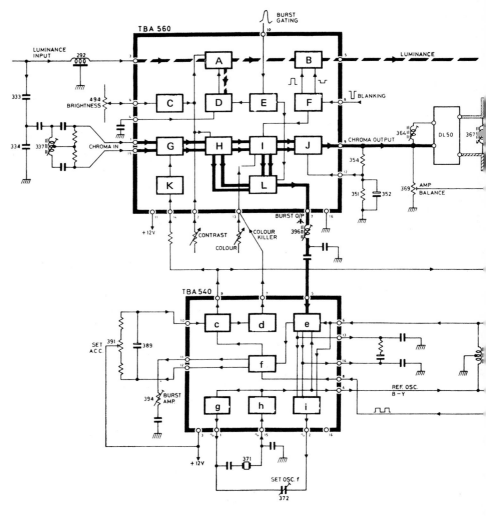

(see page 127)—which are separately amplified by transistor stages for application to the red, green and blue guns of the picture tube.

The TBA540 i.c. deals mainly with the burst operations and reference signal generation. The usual crystal oscillator is employed here and phase control is by the varicap.

The associated detector section also yields 7·8 kHz ripple signal (see page 60) from the swinging bursts for synchronising the PAL switch in the TBA990. The ripple, whose amplitude varies in sympathy with the amplitude of the bursts in this model, is also applied to the automatic chroma control (a.c.c.) detector which ultimately delivers a d.c. potential suitable for controlling the gain of the chroma amplifier.

The action of the PAL switch synchronisation or 'ident' is interesting. It will be seen that a 'flip-flop' block exists in the TBA990. The 'flip-flop' is essentially

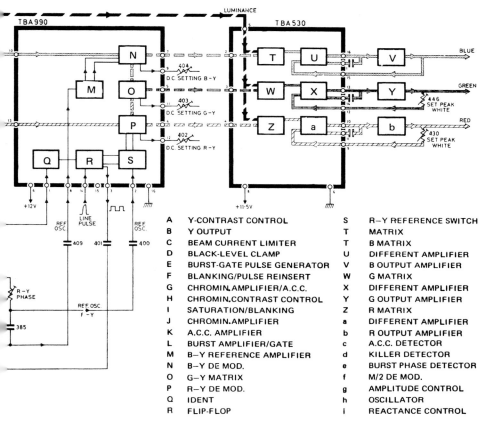

FIG. 17.3. *Simplified block diagram of the Pye 713 chassis colour circuits, showing signal paths and functions performed by four of the i.c.s used in this section (see also Fig. 17.1b).*

a bistable which is sequentially switched from one mode to the other at line frequency by line pulses applied to pin 14. This operates the R − Y reference switch, thereby reversing the phase of the reference signal, applied via C400, to the R − Y demodulator to correspond to the phase-reversed lines of chroma signal. To ensure that the reference signal phase reversals occur on the correct lines the 'ident' is necessary. The controlling potential for this is obtained from the a.c.c. detector in the TBA540.

When the PAL sync is correct the potential at pin 1 of the TBA990 is around 1 V, and under this condition the flip-flop triggers normally from the line signal. However, should the switching get in error, the control potential rises, which halts the normal triggering for one line only, because after this the sync is correct and the control potential falls to about 1 V, thereby allowing the normal, correctly synced switching to recommence. The difference in control potential is produced because the demodulator section in TBA540 receives not only the 7·8 kHz ripple pulses but also pulses from the flip-flop fed back from pin 3 of the TBA990. Error in coincidence between the two signals yields the increase in control voltage at the a.c.c. detector.

The chroma amplifier is contained in the TBA560, and the output from this is applied to the usual PAL delay line decoder such that the B — Y demodulator in the TBA990 receives only U signal, while the R — Y demodulator receives ± V signal. A separate detector in the TBA540 provides a potential for the colour killer which, of course, lies in the chroma channel of the TBA560. This operates from the demodulator in the TBA540, and is similar to the 'ident' arrangement. When there is no colour information in the signal the demodulator receives no 7·8 kHz signal (obvious, because there are then no bursts!). The input to the a.c.c. detector thus falls. This is reflected as a fall in potential from the killer detector at pin 7. The chroma section terminated to pin 13 of the TBA560 is biased-off at 0–1 V, and this occurs on monochrome since then the potential at pin 7 of the TBA540 falls to less than 1 V. On colour signal the chroma section is biased-on because pin 7 potential then rises.

The BRC 8000 chassis also uses a decoder i.c. which contains the PAL switch, the R — Y and B — Y detectors, the green matrix and matrix amplifiers which receive the luminance signal plus the three colour-difference signals and deliver for separate transistor amplification the red, green and blue primary colour signals. These are finally fed to the red, green and blue cathodes of the picture tube.

It is quite common practice nowadays to employ a separate single transistor for each primary colour output stage.

PASSIVE REFERENCE SIGNAL GENERATOR

Another scheme which is coming into more common use is the so-called 'passive' reference signal generator, and one British firm using the idea is Rank-Bush-Murphy in the recent single-standard chassis. The term 'passive' implies that there is no actual regenerative oscillator as such and that the

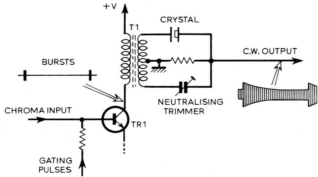

FIG. 17.4. *Basic passive subcarrier regenerator circuit. The crystal serves as a high-Q bandpass filter which removes the side frequencies from the bursts to provide a 4·43 MHz continuous wave output.*

gated-out bursts are made into a continuous wave signal by the action of the crystal.

The bursts can be regarded as a 4·43 MHz carrier-wave with sidebands either side at half the line frequency and multiples thereof. Thus, by feeding the bursts through a very narrow passband filter all the sidebands are removed and the original carrier wave only results. The narrow passband filter is, in fact, the quartz crystal, and the basic operation is illustrated in Fig. 17.4. Tr1 is the gating amplifier, and the bursts at the collector are developed across the primary of transformer T1. The secondary of T1 is loaded into the crystal circuit, this being akin to a bridge, the crystal comprising one arm and a trimming capacitor the other. The capacitor is adjusted to neutralise the capacitance of the crystal, this being necessary to 'sharpen' the tuning as much as possible.

However, in spite of this 'sharpness', the selectivity is not always great enough to attenuate adequately the half line-frequency adjacent sideband components. Thus some receivers adopting the passive reference generator type of circuit include an extra circuit which changes the swinging bursts into non-swinging NTSC counterparts.

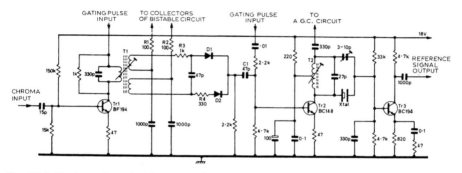

FIG. 17.5. *The burst channel with passive subcarrier generator employed in the RBM single-standard chassis.*

The RBM chassis is so engineered. From the circuit section in Fig. 17.5, it can be seen that the composite chroma signal is fed to Tr1 base, the first gating amplifier stage. The collector of this transistor is fed with clipped, positive-going line flyback pulses, and since these occur only during the bursts the transistor is energised only when bursts are present.

We have seen that the bursts swing in phase by ±45 degrees with respect to the − U axis (see page 198 and Fig. 10.20) line-by-line. It is the purpose of D1 and D2 to remove these swings and thus convert the bursts to a constant phase. Transformer T1 provides the phasing in conjunction with the alternate switching of D1 and D2. The diode switching is via pulses from the bistable PAL switching circuit. On one line, for example, the potential at R1 is positive so that D1 is switched on via the top secondary winding of T1, so the burst at the end of that line is communicated to Tr2 base through C1. D2 is switched off during this time. On the next line, however, D1 switches off and D2 switches on. The burst at the end of that line thus emanates from the bottom secondary winding. The two windings are arranged so that the phase of the burst at the end of one line is shifted 90 degrees relative to the phase of the burst at the end of the next line, as shown in Fig. 17.6. Having in mind that the bursts are swinging in phase at the same time, it will be appreciated that the 90 degree switching tends to neutralise the ± 45 degree (total 90 degrees) swings. The operation is revealed by vectors in Fig. 17.7. Notice that the 90 degree phase shift occurs on the odd lines so that the burst phase at Tr2 base settles at a constant 225 degrees.

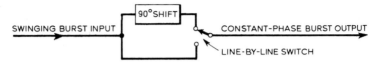

FIG. 17.6. *D1/D2 and T1 in Fig. 17.5 provide a 90 degree phase-switching action to remove the alternate line phase swings from the bursts as shown in simplified form here.*

Tr2 in Fig. 17.5 is the second gate/amplifier and the pulses gating this one are derived from an overshoot from a suitably damped tuned circuit triggered by line flyback pulses. This scheme times the gating pulses accurately.

Tr2 collector is loaded by the crystal filter, already explained (Fig. 17.4). The phase-constant reference signal (now forming the subcarrier) is amplified and limited by Tr3 prior to being processed for application to the U and V chroma detectors.

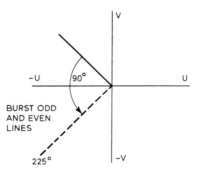

FIG. 17.7. *When the 90 degrees switch is corresponding to the bursts the phase settles at 225 degrees on all lines. If the switching is asynchronous the bursts on alternate lines cancel so that there is no reference signal. This feature is used for PAL line ident in the RBM chassis.*

The PAL switching action of this chassis is obtained by the bistable-operated switch reversing the phase of the V-chroma signal applied to the $R - Y$ detector on alternate lines. The 'ident' circuit also differs from convention to fit in with the other parameters of the decoding function. Actually, in this chassis incorrect PAL switching results in muting of the reference signal, the signal occurring only when the PAL switch is in step with the V-chroma phase alternations.

SILICON CONTROLLED SWITCH FIELD TIMEBASE

The timebase circuits have also been very much updated to cater for the solidstate age, and one now often finds in the field timebase a circuit incorporating a silicon controlled switch. The Pye 713 chassis and the Philips G8 chassis are but two examples in which this device is used.

The field timebase circuit of the Philips G8 is given in Fig. 17.8, and the

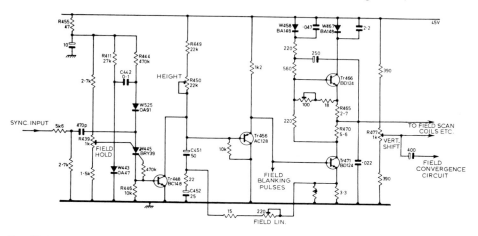

FIG. 17.8. *Field timebase circuit used in the Philips G8 chassis. A silicon controlled-switch (W445) acts as the oscillator in conjunction with discharging transistor Tr448.*

circuit operation is as follows. During the scanning stroke Tr448 is without switch-on bias and is thus non-conducting. The silicon controlled switch (s.c.s.) W445 is also non-conducting. C442 charges from the 45 V supply line through R444 and R455. As C442 charges the potential at W445 anode rises, and when this exceeds the potential at its anode gate (as established by the hold control) the device immediately 'fires' and heavy conduction results through R446. The voltage thus developed across this resistor switches on Tr448 and C451/C452 are swiftly discharged, this giving the retrace action. During the scanning stroke these two capacitors charge from the 45 V supply through R449 and the height control R450, the charging waveform effectively providing the drive for the output stage. The switching or firing of the s.c.s. is instigated at the precise moment for sync due to the application of sync pulses at its anode.

Tr456 is an emitter-follower which drives the push-pull output transistors Tr466/Tr471. The output stage, however, is in the form of a single-ended circuit, the scanning current being extracted from R465/R470 junction. The scanning coils are returned to chassis through the raster-correcting circuits, the convergence circuits and the vertical shift control R477, which regulates the amount of d.c. in the field coils.

Diode D525 in the s.c.s. circuit is included to prevent the positive-going field sync pulses from being shorted to chassis through C442 and W443.

LINE TIMEBASES

In the main these are updated versions of earlier-developed circuits. The transistorised flywheel-controlled sine wave type of oscillator circuit is commonplace. Line output stages are now getting away with a single power transistor, and the circuitry involved is not all that much different from that employed in the early all-transistor BRC models (see pages 105–107). However, the circuits appear to be engineered towards greater reliability in this rather vulnerable area.

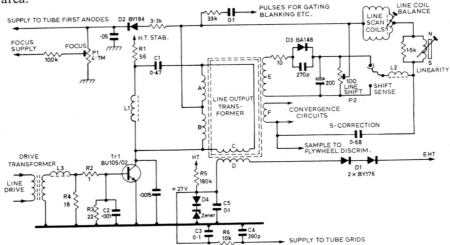

FIG. 17.9. *Solidstate line output circuit of the BRC 8000 chassis.*

The circuit of the line output stage of the BRC 8000 chassis is given in Fig. 17.9. A positive supply for the collector of the npn output transistor is obtained from a stabilised 180 V source, and is fed through R1/L1. Output transformer primary consists of windings A and B, and because the h.t. side of L1 is connected through C1 to the top of winding A while the bottom of winding B is connected to Tr1 collector, the transformer is shunt fed, this having the advantage of removing the d.c. from the primary windings and putting it through the less vulnerable L1, which has a high impedance to line frequency signals.

Winding C in parallel with windings A and B couples the e.h.t. overwind D, which feeds flyback pulses of suitable amplitude to dual-stage e.h.t. rectifier D1, the d.c. output being in the order of 22 kV. D2 provides potential for the tube's first anode, the source also giving the focusing potential, adjustable by P1.

Winding E feeds rectifier D3 which delivers d.c. suitable for horizontal shift control (P2). The tapping to L2 changes the direction of shift if necessary. The convergence system, which is low impedance, is connected in the 'earthy' side of the scanning circuit.

Suitable drive signal (from a flywheel-controlled oscillator) is applied to Tr1 base via the transformer, and at the end of a scanning stroke the transistor switches off. The required low impedance base circuit is provided by R2/R3/R4 and C2, R2 being a current limiting resistor. L3 in the base circuit introduces a small delay so that the charge energy stored by the collector junction can be smoothly exhausted without affecting the timebase operation. R3 inhibits 'ringing' and gives a degree of protection to the output transistor in the event of a fault condition which might permanently or intermittently break the circuit from the driver transformer.

When the circuit tries to swing negatively, subsequent to the production of the e.h.t. pulse at Tr1 collector, the collector/base junction of the transistor conducts and provides an 'efficiency diode' action; that is, the 'earthy' side of the circuit is clamped to chassis and the current decay in the line coils then provides the first part of the scanning stroke. The circuit is also tuned to the third harmonic of the retrace for improved e.h.t. regulation.

It looks as though the thyristor will soon be found in more line output stages. At least one such circuit has been developed—by ITT for use with this firm's thin-neck tubes.

BEAM CURRENT LIMITING

It will be seen that the low potential side of the e.h.t. overwind is returned to chassis through D4 and the zener diode in series. These diodes are biased from the stabilised 180 V supply line through R5, and pass a current of 1 mA. This results in +27V being present at D4 anode, and after filtering by C3/R6/C4 this potential is applied to the three grids of the picture tube (the primary colour drive signals being applied to the three cathodes).

Now, should the tube beam current exceed about 1 mA the diodes cease to conduct and the voltage at D4 anode goes less positive since then the negative potential at the low potential side of the e.h.t. overwind outweighs more of the positive potential from the 180 V supply line, via R5. This back-biases the tube guns and thus reduces the beam current.

The charges of C3/C4 are retained for a short period after the receiver is switched off, and since the tube grids are then more positive than the cathodes a heavy beam current flows which almost instantaneously discharges the e.h.t. reservoir capacitor, thereby avoiding the lingering switch-off spot in the centre of the screen.

OTHER SUPPLIES FROM LINE SCAN ENERGY

Line scan energy in the output stage is increasingly being used to power other sections of the receiver. Separate windings on the line output transformer are connected to rectifiers and the resulting d.c. used for the powering. In the Pye 713 chassis, for example, supplies for transistors and even the heaters of the picture tube are obtained this way. The main supply for the line output stage is obtained from a stabilising circuit using two thyristors and a transistor (no mains transformer). To get the line timebase working in the first place, however,

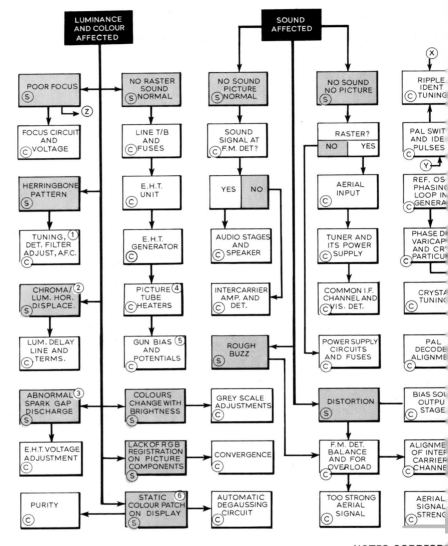

NOTES CORRESP[...]

1. A.F.C. fault can cause detuning and sound/chroma be[...]
2. This is different from misconvergence.
3. Sometimes indicates too high e.h.t. voltage.
4. Usually possible to see these glowing.
5. Brightness control circuit fault could bias-off all guns[...]
6. Can be caused by hi-fi speaker too close to receiver!
7. Check ref. osc. signal phasing in particular.
8. Class B push-pull requires correct quiescent current [...]
   least distortion.
9. For example, aerial of incorrect channel group.
10. Also caused by inadequate aerial signal or advanced [...]
    setting of colour control.

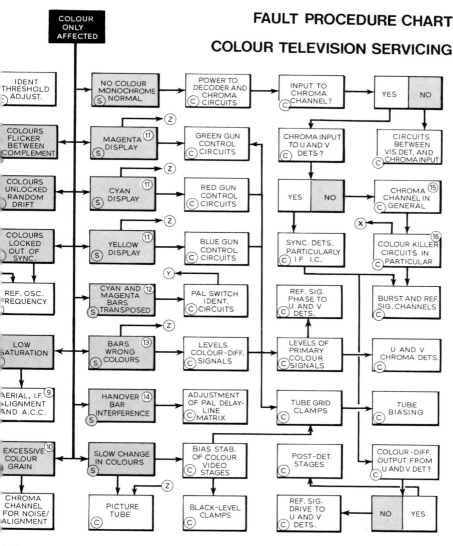

UMBERS IN BLOCKS

Complementary colour displays indicating failure of one beam. Failure of two beams (unusual) yields displays of active beam primary colour.

Usually signifying PAL switch count in error.

Symptom depends on exact area of fault and on whether colour-difference or primary colour drive employed. Also check colour drive and gain presets.

Co-channel interference another cause. Higher gain aerial might be necessary.

Check signal through chroma channel with oscilloscope.

Over-riding colour killer often helps.

(C) = CHECK

(S) = SYMPTOM

(X)
(Y) } = ASSOCIATIONS
(Z)

the main supply is dropped resistively and then coupled to the line oscillator via a diode switch. Once the rectified supplies from the line output transformer become available the diode switches off and the main supply is disconnected.

The thyristor method of main supply stabilisation is interesting, and is used in several contemporary models with differences mostly in circuit detail. It will be recalled that valve receivers were equipped with line timebase stabilisation (i.e., a voltage-dependent resistor sampling line pulses and producing a control potential for the output valve's control grid of amplitude depending on the line output stage loading and hence on the beam current, a kind of line output a.g.c.—see Fig. 7.9 on page 102). This kind of stabilisation is inapplicable to transistor line timebases, so the power supply proper needs to be stabilised. One or two early transistor receivers featured e.h.t. regulation (see page 108, for instance) and h.t. stabilisation, but the current trend is towards a well stabilised main power supply which reflects into the other circuits.

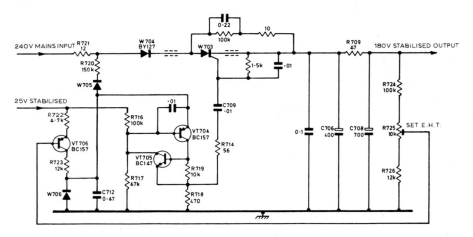

FIG. 17.10. *Power supply regulation in this circuit from the BRC 8000 chassis is based on the thyristor. See text for details of operation.*

Fig. 17.10 gives the circuit of the stabilised supply of the BRC 8000 chassis. This yields a nominal 180 V output from an input of 240 V 50 Hz (mains supply) and has an output current range of 0–600 mA from an input current of 1·7 A. Effective source resistance is less than 10 ohms over the full current range, and the stabilisation is such that there is less than 3 V change over 200–270 V input. The circuit calls for an extra 25 V stabilised supply at a mere 1·5 mA, which is not difficult to obtain (see later).

Operation is based on thyristor W703 between the input and output and rectifier W704 on the mains input side conducting on the positive swings. The thyristor acts as a controlling switch which is triggered into conduction by positive pulses applied to its gate during the positive swings of the mains supply for a period governed by the input voltage and the load demands. In other words, the timing of the triggering pulses controls the output voltage.

VT706 base is in communication with the output voltage picked up from the divider R724/R725/R726, the base thus receiving a voltage determined by the setting of R725 and proportional to the output voltage. Now, since VT706 emitter is returned to the stabilised 25 V reference source through R722, the transistor's emitter and hence collector current varies as a function of the output voltage.

Assuming that the supply is swinging negatively, diode W705 goes into forward bias, so that any charge in C712 and VT706 collector current are returned to the mains supply through R720. Thus the voltage on C712 will be drawn negatively until diode W706 conducts and limits the voltage across C712 to just below zero.

When the supply swings positively W705 goes into reverse bias and switches off, thereby allowing VT706 collector current to commence charging C712. A ramp voltage thus develops across C712, which also appears at VT704 emitter. The rate of rise of this ramp voltage is dependent on the value of the regulated output voltage.

VT704 base is held at about 10 V by divider R716/R717 so that the base/emitter junction is reverse-biased. When the ramp voltage exceeds this value the transistor commences to conduct. When this happens its collector current forward-biases VT705 base/emitter junction, via R719/R718, so that this transistor also conducts. The effect is regenerative owing to the coupling from VT705 collector to VT704 base, which means that the two transistors turn on hard very swiftly.

The action rapidly discharges C712 through the transistor pair and R718. Thus as soon as the ramp voltage reaches the transistor switching value, determined by R716/R717, C712 discharges. The resulting sharp pulse of the rapid discharge, which appears across R718, is coupled by C709 and limiting resistor R714 to the thyristor gate, so that this device is immediately turned on.

The reservoir C706 then charges from the mains supply via limiting resistor R721 and W704 until the mains voltage falls to meet the rising voltage across C706, at which time the thyristor turns off. Supply smoothing is by R709/C704.

As the mains swings negatively again C712 is discharged through W705/R720, and VT704 base/emitter junction is again reverse-biased, ready for the next positive half-cycle.

The time at which the thyristor is triggered thus varies in accordance with the *rate of rise* of the ramp voltage, which itself is a function of the output voltage as we have seen. Thus, if the output voltage tends to fall due to a rise in load current, for example, the thyristor is triggered earlier during the positive half-cycle to restore the output voltage to the correct value.

D704 serves essentially as a protection device for the thyristor during negative swings of the mains supply, and thus allows a thyristor which is incapable of withstanding high reverse voltage to be used.

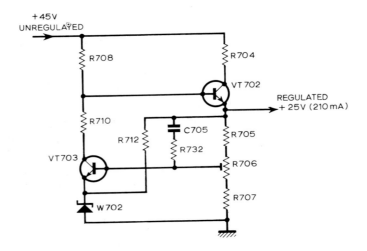

FIG. 17.11. *Simple series regulator also used in the BRC 8000 chassis. This provides the 25 V reference potential required for the circuit in Fig. 17.10 and also power for other parts of the receiver.*

SIMPLE SERIES REGULATOR

Less complex series regulators are also found in some of the recent all-transistor receivers, and that shown in the circuit in Fig. 17.11 is a fairly typical example. It is this circuit, in fact, which provides the 25 V stabilised source for the thyristor control (Fig. 17.10) of the BRC 8000 chassis. The power is well in excess of the 1·5 mA reference demands since the source supplies other sections of the receiver as well.

The series regulator transistor is VT702 between the 45 V unstabilised input and the 25 V stabilised output. This is driven conventionally with the correct voltages by VT703 and W702 as the reference. C705/R732 increase the loop gain at 100 Hz for maximum ripple reduction. Output voltage is set by R706.

OTHER TRENDS

Printed circuit i.f. transformers are also being used, while many front-ends (u.h.f. tuners) incorporate varicaps for quarter-wave line tuning. Automatic frequency control (a.f.c.) is also commonplace, the control potential being obtained from a discriminator circuit tuned to the i.f. and picking up signal from the main i.f. channel.

Not all models are keeping to e.h.t. triplers. The BRC 8000, for example, obtains its e.h.t. voltage from an overwind of fairly high pulse voltage on the line output transformer, as already shown (Fig. 17.9). Fault vulnerability due to the 22 kV overwind, however, is not as great as may be imagined because the transformer is subjected to silicon encapsulation and because there are only two rectifiers in the circuit, as compared with a possible five (plus four capacitors) of a tripler assembly. The Pye 713 chassis employs an encapsulated e.h.t. doubler.

Integrated circuits are now commonly used in the sound intercarrier/f.m. detector sections. Some f.m. sections, in fact, are of the so-called 'slope' type. Low-level synchronous switching detectors are also being used for vision. Such an arrangement is incorporated in the BRC 8000 chassis, the detector with the final i.f. stage being contained in an i.c.

Shadowmask tubes with 110 degree deflection angle are appearing in a few models at the time of writing. The principle of operation remains unchanged, but the wider angle of deflection poses some additional problems with regard to pin-cushion distortion correction and dynamic convergence, particularly at the corners of the screen. Various circuits, some quite complicated in imported receivers, have been devised to overcome the problems and the current trend here is towards simplification of the controlling circuits. At the time of writing, however, there is no 'standard' circuitry that can be discussed; but all up-to-date information will ultimately appear in the complementary *Newnes Colour TV Servicing Manual*.

In general, servicing is calling for greater use of the oscilloscope, and this applies to speedy fault diagnosis in the customer's home as well as in the workshop, even if the former merely leads to replacement of a printed circuit board or module and later repair in the workshop. A good multirange test meter also remains a key tool, and a great deal of time can be saved with a properly coded colour pattern generator, although many technicians are finding that quite a few servicing operations can be undertaken with off-air signals.

CHAPTER 18

Television Standards

TABLE 18.1 gives the television standards used in different parts of the world. The code letters are those of the CCIR, and Table 18.2 relates these to the various primary parameters. With the exception of the 'M' system of the USA and some other countries where 60 Hz is the standard, the field frequency is 50 Hz.

Of particular interest are the fairly recent Australian, New Zealand and

Table 18.1   World television standards

Aden (B)	Gibraltar (B)	Nicarague (M)
Albania (D)	Guatemala (M)	Nigeria (B)
Algeria (B, E)***	Haiti (M)	Norway (B)*
Antigua (M)	Hawaii (M)	Pakistan (B)
Argentina (N)	Honduras Republic (M)	Panama (M)
Australia (B)*	Hong Kong (A, I)*	Peru (M)
Austria (B, G)*	Hungary (D)	Philippines (M)
Barbados (N)	Iceland (B)	Poland (D)
Belgium (C, G, H)*	India (B)	Portugal (B, G)
Bermuda (M)	Iran (B)	Puerto Rico (M)
Brazil (M)	Iraq (B)	Rhodesia (B)
Bulgaria (D)	Israel (B)	Rumania (D)
Cambodia (M)	Italy (B, G)	Ryukyu Islands (M)
Canada (M)	Ivory Coast (E)	Saudia Arabia (M, B)*
China (D)	Jamaica (N)	Sierra Leone (B)
Colombia (M)	Japan (M)	South Africa (I)*
Congo (B)	Kenya (B)	Spain (B, G)
Costa Rica (M)	Korea (M)	Sweden (B, G)*
Cuba (M)	Kuwait (B)*	Switzerland (B, G)*
Cyprus (B)	Lebanon (B)**	Syria (B)
Czechoslovakia (D)	Liberia (B)	Thailand (M)
Denmark (B, G)*	Libya (G)	Trinidad (M)
Dominican Republic (M)	Luxembourg (F, L)*	Tobago (M)
Ecuador (M)	Malaysia (B)	Tunisia (B)
Egypt (B)	Malta (B)	Turkey (B)
Eire (A, I)*	Marianas Islands (M)	UK (A, I)
El Salvador (M)	Mauritius (B)	USA (M)
Ethiopia (B)	Mexico (M)	USSR (D)
Finland (B, G)*	Monaco (E)	United Arab Republic (B)
France (E, L)**	Morocco (B)	Uruguay (M)
Germany, West (B, G)*	Netherlands (G, B)*	Venezuela (N)
Germany, East (B)	Nether-Antilles (B)	Yugoslavia (B, H)*
Ghana (B)	New Zealand (B)*	Zambia (B)

* PAL system.   ** SECAM system.   *** Proposed SECAM system.

South African systems, where the PAL colour system is in use or will be adopted. To help readers locate the points of difference between the receivers used in these countries and those used in the UK on the 625-line system 'I', each of the above mentioned countries are considered separately below.

Table 18.2   Television systems

System code	No. of lines	Channel width MHz	Vision band-width MHz	Sound separation (from vision)	Vestigial side-band MHz	Vision mod. sense	Sound mod.
A	405	5	3	− 3·5	0·75	+	a.m.
B	625	7	5	+ 5·5	0·75	−	f.m.
C	625	7	5	+ 5·5	0·75	+	a.m.
D	625	8	6	+ 6·5	0·75	−	f.m.
E	819	14	10	±11·15	2	+	a.m.
F	819	7	5	+ 5·5	0·75	+	a.m.
G	625	8	5	+ 5·5	0·75	−	f.m.
H	625	8	5	+ 5·5	1·25	−	f.m.
I	625	8	5·5	+ 6	1·25	−	f.m.
K	625	8	6	+ 6·5	0·75	−	f.m.
L	625	8	6	+ 6·5	1·25	+	a.m.
M	525	6	4·2	+ 4·5	0·75	−	f.m.
N	625	6	4·2	+ 4·5	0·75	−	f.m.

Some of the primary points of difference lie in the i.f., intercarrier and rejector tuning. With UK system 'I' receivers the vision and sound i.f.s. are respectively in the order of 39·5 MHz and 33·5 MHz, with the adjacent vision and adjacent sound rejectors tuned round 31·5 MHz and 41·5 MHz respectively, and with the intercarrier at 6 MHz.

AUSTRALIAN SYSTEM

This country uses the 'B' system currently on channels 0–11 (from 46·25 MHz Channel 0 vision to 221·75 MHz Channel 11 sound). The ratio of effective radiated power sound/vision is 20 per cent, the field frequency 50 Hz and the mains supply 240 V.

Currently used i.f.s are vision 36 MHz and sound 30·5 MHz, with adjacent vision rejection at 29 MHz and adjacent sound rejection at 37·5 MHz.

To avoid Channel 2 beat patterns alternative frequencies have been proposed. These, which *might* be adopted, are vision 36·875 MHz and sound 31·375 MHz, with adjacent vision rejection at 29·875 MHz and adjacent sound rejection at 38·375 MHz.

Owing to the 5·5 MHz difference between the sound and vision carriers the intercarrier frequency is 5·5 MHz, as distinct from the 6 MHz intercarrier of the UK 'I' system.

COLOUR TELEVISION SERVICING

NEW ZEALAND SYSTEM

This country also uses the 'B' system with the nominal vision carrier at 1·25 MHz above the lower frequency limit of the channel and the unmodulated sound carrier provisionally 5·5 MHz above the vision carrier, thereby giving an inter-carrier frequency of 5·5 MHz. Four Bands are available for television (i.e. Band I 41–51 MHz and 54–68 MHz; Band III 174–216 MHz; Band IV 470–585 MHz and Band V 610–690 MHz) but are not exclusively allocated to the television service in New Zealand.

Channel width is 7 MHz in the v.h.f. bands and 8 MHz in the u.h.f. bands.

Intermediate frequencies are 33·4 MHz sound, 38·9 MHz vision and 34·47 MHz (approximately) colour subcarrier (all within ±0·25 MHz). These frequencies provide the best compromise between various forms of interference that can arise from image responses, i.f. difference frequency responses, i.f. harmonic interference and beat oscillator radiation. Adjacent vision and adjacent sound rejection frequencies are in the order of 31·9 MHz and 40·4 MHz respectively.

Mains voltage is 230, 50 Hz.

Chromaticity coordinates are *red* x = 0·64 and y = 0·33; *green* x = 0·29 and y = 0·6; *blue* x = 0·15 and y = 0·06; and illuminant D6500 x = 0·313 and y = 0·329 (see Plate 1 facing page 136).

SOUTH AFRICAN SYSTEM

The 'I' system used in this country is virtually a replica of that used in the UK, but with minor alterations as detailed below.

1. Vision carrier to sound carrier power ratio 10:1.
2. Although the sound signal has a maximum deviation of ±50 kHz and is subjected to 50 μs pre-emphasis (as in the UK system), the definition of 'sound carrier modulation' is slightly different to take account of the type of peak programme meter scales used in South Africa.
3. (a) Band I is not used.
   (b) Band III is extended to 254 MHz, but Channel 12 is not used.
   (c) Bands IV/V are channelled in the same way as in the UK and Europe, but the only gap is Channel 38.

As in the UK, the u.h.f. plan makes provision for the assignment of four channels per station on a regular basis, in an arrangement of n, n + 4, n + 8 and n + 12. Thus, except for a few cases where Channel 38 needs to be omitted, leaving three u.h.f. channels only, all stations have an assigned four channels.

Television aerials will be co-sited with existing f.m. aerials, of which there are currently (1973/74) in excess of 80 f.m. stations in operation, providing a coverage in advance of 98 per cent of the population. Eighteen main TV stations are planned to come into operation at the start of the official service in January 1976, plus a number of low-power 'gap-filling' stations, and it is predicted that more than 80 per cent of the population will be covered by these stations.

Intermediate frequencies are 38·9 MHz vision and 32·9 MHz sound, the difference between these and those of UK receivers being insignificant. Adjacent vision and adjacent sound rejection frequencies are thus in the order of 30·9 MHz and 40·9 MHz respectively.

Since the u.h.f. bands only are to be used, receivers with exclusive v.h.f. channelling and tuning ranges would not be suitable for use in South Africa.

Six manufacturing groups will be allowed to supply the South African market with receivers complying with the specification published by the South African Bureau of Standards, and these will represent most of the major overseas manufacturers. However, the receivers marketed will not necessarily be identical with models of the same firm marketed in different countries.

Acknowledgements are given to the South African Broadcasting Corporation for the help provided in obtaining the information presented herewith under the South African section.

I would also like to convey my thanks to Mr. R. V. Arnaboldi of Thorn Consumer Electronics for his assistance in obtaining information on receivers for overseas use. It is this firm's intention so far as Australia and South Africa are concerned to introduce a new chassis known as the 4000 series which incorporates 110 degree tube technology; virtually the whole production of this chassis is destined for exportation. Local technicians and engineers will introduce the modifications to the chassis to meet the particular market requirements, and in the first instance most of the components involved will be shipped from the UK factories.

With regard to New Zealand, initial Thorn Group shipments are of the UK 3500 series 90 degree chassis components and assemblies; though in 1974 the Company's New Zealand plant will be changed over to a version of the 4000 series.

Other major British manufacturers will also be contributing to the requirements of the television-developing countries, so much of the information presented in this book will have value to service technicians in those countries as well as to those operating in the UK. Clearly, there will be modifications, particularly to the tuned circuits, but these can be defined from the foregoing information.

Index